Principles of Chemical Separations with Environmental Applications

Chemical separations are of central importance in many areas of environmental science, whether it is the clean-up of polluted water or soil, the treatment of discharge streams from chemical processes, or modification of a specific process to decrease its environmental impact. This book is an introduction to chemical separations, focusing on their use in environmental applications.

The authors first discuss the general aspects of separations technology as a unit operation. They also describe how property differences are used to generate separations, the use of separating agents, and the selection criteria for particular separation techniques. The general approach for each technology is to present the chemical and/or physical basis for the process, and explain how to evaluate it for design and analysis.

The book contains many worked examples and homework problems. It is an ideal textbook for undergraduate and graduate students taking courses on environmental separations or environmental engineering.

RICHARD NOBLE received his Ph.D. from the University of California, Davis. He is a professor of Chemical Engineering and Co-Director of the Membrane Applied Science and Technology Center at the University of Colorado, Boulder.

PATRICIA TERRY received her Ph.D. from the University of Colorado, Boulder, and is an associate professor in the Department of Natural and Applied Sciences at the University of Wisconsin, Green Bay. She has also held positions at Dow Chemicals and Shell Research and Development.

Principles of Chemical Separations with Environmental Applications

Richard D. Noble
University of Colorado, Boulder

and

Patricia A. Terry
University of Wisconsin, Green Bay

CAMBRIDGE
UNIVERSITY PRESS

PUBLISHED BY THE PRESS SYNDICATE OF THE UNIVERSITY OF CAMBRIDGE
The Pitt Building, Trumpington Street, Cambridge, United Kingdom

CAMBRIDGE UNIVERSITY PRESS
The Edinburgh Building, Cambridge, CB2 2RU, UK
40 West 20th Street, New York, NY 10011-4211, USA
477 Williamstown Road, Port Melbourne, VIC 3207, Australia
Ruiz de Alarcón 13, 28014 Madrid, Spain
Dock House, The Waterfront, Cape Town 8001, South Africa

http://www.cambridge.org

First published 2004

Printed in the United Kingdom at the University Press, Cambridge

Typefaces Times 10/14 pt and Gill Sans   *System* LaTeX 2_ε [TB]

A catalogue record for this book is available from the British Library

Library of Congress Cataloguing in Publication data

Noble, R. D. (Richard D.), 1946–
Principles of chemical separations with environmental applications / Richard D. Noble and Patricia A. Terry.
      p.   cm. – (Cambridge series in chemical engineering)
Includes bibliographical references and index.
ISBN 0 521 81152 X – ISBN 0 521 01014 4 (pbk.)
1. Separation (Technology)   2. Environmental chemistry.   3. Environmental management.
I. Terry, Patricia A. (Patricia Ann), 1965–   II. Title.   III. Series.
TP156.S45N63   2003
660′.2842–dc21   2003053072

ISBN 0 521 81152 X hardback
ISBN 0 521 01014 4 paperback

The publisher has used its best endeavors to ensure that the URLs for external websites referred to in this
book are correct and active at the time of going to press. However, the publisher has no responsibility
for the websites and can make no guarantee that a site will remain live or that the content is or
will remain appropriate.

Contents

Contents

Preface

Separation – the process of separating one or more constituents out from a mixture – is a critical component of almost every facet of chemicals in our environment, whether it is remediation of existing polluted water or soil, treatment of effluents from existing chemical processes to minimize discharges to the environment, or modifications to chemical processes to reduce or eliminate the environmental impact (chemically benign processing). Having said this, there is no text today for this subject which describes conventional processing approaches (extraction, ion exchange, etc.) as well as newer techniques (membranes) to attack the serious environmental problems that cannot be adequately treated with conventional approaches. Existing texts for this subject primarily focus on wastewater treatment using technology that will not be suitable in the larger context of environmental separations. Interestingly, most chemical engineering texts on separations technology are primarily based on whether the separation is equilibrium or rate based. Thus, it is difficult to find one source for separations technology in general.

This text is meant as an introduction to chemical separations in general and various specific separations technologies. In Chapter 1 we give a generalized definition of separation processes and their environmental applications. Following this, the approach to the organization of this text is to first discuss, in Chapter 2, the generic aspects of separations technology as unit operations. This chapter will include a discussion of the use of property differences to generate the separation, the use of a separating agent to facilitate the separation, as well as some discussion on the criteria for selection of a particular separation process. This last point is usually discussed at the end of a text on separations, but we felt that it was better to give students this "food for thought" prior to any description of specific technologies.

Mass transfer fundamentals, including equilibrium- and rate-based mechanisms, are introduced in Chapter 3, before any description of specific technologies. Many readers will be chemists, civil engineers and others with little or no previous experience in the design or analysis of these processes. It is important that everyone be "brought up to speed" prior to any discussion of a specific process. If this is not done, each technology appears to have its own set of rules and design algorithms. This "unique" set for each process diminishes the ability of the reader to use generic principles to compare alternatives and evaluate new approaches as they become available. Once this major division of the approaches has been covered, later chapters describe the specific technologies.

The section in Chapter 3 on equilibrium stage separations will include both graphical and analytical techniques. The graphical techniques are useful to visualize the process for the student and the analytical methods reinforce the principles. Rate-based separations will focus on diffusional processes and convective/dispersive effects which can be described by mass transfer coefficients (k). Initial discussion will focus on which approach to use based on what information is available and what one wants to determine. For analyses using mass transfer coefficients, both the use of correlations to estimate a value for k and the determination of an overall mass transfer coefficient (K) will be covered.

In discussing individual separations technologies in Chapters 4 through 9 we consider separations using physical property differences as well as chemical interactions. Distillation, extraction, absorption, adsorption, ion exchange, and membranes are covered. Our approach to each technology is not to provide an exhaustive description. Rather, we want to explain the physical and/or chemical basis for the process and how to evaluate it for design or analysis. Books that describe a given technology in detail will be given as references. Membrane separations represent a new and emerging technology which has been used commercially for filtration and gas separation. It is a topic that is rarely discussed in any text on separations, so we plan to insure that it receives adequate coverage.

Special thanks go to the students that assisted us including Kendra Axness, Katie Benko, Liz Galli, Jill Gruber, Blue Parish, Laura Weber, and Tony Worsham. We also want to thank others in the chemical separations community that helped to encourage us along the way including Ed Cussler, Phil Wankat, Jud King, Ed Lightfoot, Norman Li and Bill Koros. I (RDN) would like to thank Ben McCoy who taught my first separations class and started me, perhaps inadvertently, on this career path.

We are deeply indebted to Ellen Romig. Without her help in the typing and editing, it is highly doubtful that this book would have seen the light of day.

1

Introduction

When the well's dry, we know the worth of water.

– BENJAMIN FRANKLIN (1706–1790), *Poor Richard's Almanac*, 1746

You can't always get what you want, but, if you try sometimes, you get what you need.

– ROLLING STONES, 1969

1.1 Objectives

1 Define separation processes and explain their importance to environmental applications.
2 Describe equilibrium- and rate-based analysis of separation processes.
3 List pollution sources for water, air, and soil.
4 Give examples of clean-up of existing pollution problems and pollution prevention.
5 Describe the hierarchy of pollution prevention.
6 Discuss the relationship between degree of dilution and cost of separations.
7 Be able to state the three primary functions of separation processes.

1.2 Why study environmental applications?

The National Research Council released a report [1] that states:

The expanding world population is having a tremendous impact on our ecosystem, since the environment must ultimately accommodate all human-derived waste materials. The industries that provide us with food, energy and shelter also introduce pollutants into the air, water, and land. The potential for an increasing environmental impact will inevitably result in society's setting even lower allowable levels for pollutants.

Table 1.1 *US Environmental Industry segments [2].*

Services	Resources	Equipment
Consulting and Engineering	Water Utilities	Water Equipment and Chemical
Waste Management	Energy Sources and	Instruments and Information Systems
• Solid waste	Recovery	Air Pollution Control Equipment
• Hazardous waste	Resource Recovery	Waste Management Equipment
• Water		Process and Prevention Technology
Remediation		
Industrial Services		
Analytical Services		

This material is used by permission of Environmental Business International, Inc.

Table 1.2 *The Environmental Industry in the United States in 1992 [4].*

Sector	Approximate size	Approximate growth
Engineering and Consulting	$ 12 billion	15% over 10 years
Water Supply and Treatment	$ 30 billion	5%
Air Quality	$ 6 billion	15%
Equipment/New Technology	$ 11 billion	N/A

The report further concludes, "In the future, separation processes will be critical for environmental remediation and protection."

Chemical separations are used to reduce the quantity of potentially toxic or hazardous materials discharged to the environment. In addition, separations that lead to recovery, recycle, or reuse of materials also prevent discharge.

The US Environmental Industry is made up of many segments. Table 1.1 lists the major segments and their chief components [2]. It is apparent that chemical separations play a large role in each of these areas. In addition, processes to separate and purify chemicals consume over 10^{15} BTU of energy (BTU = 1,055 joules) alone in the United States each year. They directly or indirectly generate considerable emissions, which pose challenges that will require new processing approaches [3].

The Environmental Industry in the US is large and projected to grow at a substantial rate. Table 1.2 provides some data related to environmental applications of separations. Even if the projections are "overly enthusiastic," it is clear that this is an important technology area and will continue to grow.

1.3 Background

The topic of the material in this text is chemical separations with environmental applications. Separation processes are any set of operations that separate solutions of two or more components into two or more products that differ in composition. These may either remove a single component from a mixture or separate a solution into its almost pure components. This is achieved by exploiting chemical and physical property differences between the substances through the use of a separating agent (mass or energy).

Separation processes are used for three primary functions: purification, concentration, and fractionation. Purification is the removal of undesired components in a feed mixture from the desired species. For example, acid gases, such as sulfur dioxide and nitrogen oxides, must be removed from power plant combustion gas effluents before being discharged into the atmosphere. Concentration is performed to obtain a higher proportion of desired components that are initially dilute in a feed stream. An example is the concentration of metals present in an electroplating process by removal of water. This separation allows one to recycle the metals back to the electroplating process rather than discharge them to the environment. Lastly, in fractionation, a feed stream of two or more components is segregated into product streams of different components, typically relatively pure streams of each component. The separation of radioactive waste with short half-lives from that having much longer half-lives facilitates proper handling and storage.

Analysis of separation processes can be placed into two fundamental categories: equilibrium-based and rate-based processes. These separation categories are designated using thermodynamic equilibrium relationships between phases and the rate of transfer of a species from one phase into another, respectively. The choice of which analysis to apply is governed by which is the limiting step. If mass transfer is rapid, such that equilibrium is quickly approached, then the separation is equilibrium limited. On the other hand, if mass transfer is slow, such that equilibrium is not quickly approached, the separation is mass transfer limited. In some separations, the choice of analysis depends upon the type of process equipment used.

Equilibrium processes are those in which cascades of individual units, called stages, are operated with two streams typically flowing countercurrent to each other. The degree of separation in each stage is governed by a thermodynamic equilibrium relationship between the phases. One example is distillation, in which a different temperature at each stage alters the vapor-phase equilibrium between a typically binary mixture. The driving force for separation is the desire of a new equilibrium between the two phases at the temperature of each stage. The end result is the separation of two liquids with dissimilar boiling temperatures. Other equilibrium-based processes that will be covered in this text include extraction and solid extraction, or leaching. Extraction is the removal of a species from a liquid in which it is dissolved by means of another liquid for which it has a higher affinity, and leaching is the removal of a species from a solid phase by means of a liquid for which it has stronger affinity.

Rate-based processes are limited by the rate of mass transfer of individual components from one phase into another under the influence of physical stimuli. Concentration gradients are the most common stimuli, but temperature, pressure, or external force fields can also cause mass transfer. One mass transfer based process is gas absorption, a process by which a vapor is removed from its mixture with an inert gas by means of a liquid in which the vapor is soluble. Desorption, or stripping, on the other hand, is the removal of a volatile gas from a liquid by means of a gas in which the volatile gas is soluble. Adsorption consists of the removal of a species from a fluid stream by means of a solid adsorbent with which the species has a higher affinity. Ion exchange is similar to adsorption, except that the species removed from solution is replaced with a species from the solid resin matrix so that electroneutrality is maintained. Lastly, membrane separations are based upon differences in permeability (transport through the membrane) between components of a feed stream due to size and chemical selectivity for the membrane material.

1.4   Pollution sources

Sources of pollution vary from small-scale businesses, such as dry cleaners and gas stations, to very large-scale operations, such as power plants and petrochemical facilities. The effluent streams of industry are particularly noticeable because of their large volumes [1]. Sources include both point-source and non-point-source pollution. Point-source pollution can be traced directly to single outlet points, such as a pipe releasing into a waterway. Non-point-source pollutants, on the other hand, such as agricultural run-off, cannot be traced to a single definite source. The emissions from both span a wide range of gas, liquid, and solid compounds.

A large majority of air-polluting emissions come from mobile sources. The automobile is an obvious example, but other vehicles, such as trucks, trains, and aircraft also contribute. Emissions from mobile sources include CO_2, volatile organic compounds (VOCs), NO_x, and particulates. The last may also have heavy metals, such as lead or mercury, or hazardous organics attached. Stationary sources typically burn or produce fossil fuels – coal, gasolines, and natural gas. This produces gaseous sulfur compounds (H_2S, SO_2, etc.), nitrogen oxides (NO_x), CO_2 and particulates. Fuel producers and distributors also typically produce VOCs. Most of these pose human health concerns and many contribute to the acid-rain problem and global warming effect.

Water pollution also comes from a variety of sources. Agricultural chemicals (fertilizers, pesticides, herbicides) find their way into groundwater and surface water due to water run-off from farming areas. For example, agricultural drainage water with high concentrations of selenium threatens the Kesterson National Wildlife Refuge in California. Chemical discharge from sources ranging from household releases (lawn fertilizers, detergents, motor oil) to industrial releases into surface or groundwater supplies is an obvious problem. Industrial discharges can occur due to leaking storage facilities as well as process effluent. Municipal water treatment effluent is another prevalent source. MTBE, a gasoline additive used until recently to reduce air pollution, has been identified as a source of water pollution, demonstrating that the solution to one environmental concern can create a problem elsewhere. Isolation and recovery of these and other water pollutants pose a challenge to develop innovative separation techniques.

Pollution of soils also occurs through a variety of sources. Municipal and industrial waste has been buried in landfills, which sometimes leak, even if lined with durable impermeable materials. Periodic news accounts of hazardous chemicals migrating through soil to threaten water supplies and homes are reminders of this issue. Chemical discharge directly onto surface soil from periodic equipment cleaning, accidental discharges (spills), abandoned process facilities or disposal sites is another environmental challenge. Subsurface contamination can also occur as a result of leaking underground storage tanks.

In addition to air, water, and soil pollution, large quantities of solid and liquid wastes generated by both industry and domestic use must be remediated, recycled, or contained. Industrial wastes include overburden and tailings from mining, milling, and refining, as well as residues from coal-fired steam plants and the wastes from many manufacturing processes. The nuclear and medical industries generate radioactive solid wastes that must be carefully handled and isolated. Effective ways of fractionating long-lived radioactive isotopes from short-lived ones are needed because the long-lived ones require more expensive handling and storage. The environmental problems of residential wastes are increasing as the population grows. It is important to segregate and recycle useful materials from these wastes. In many places, there are no effective options for dealing with toxic liquid wastes. Landfill and surface impoundment are being phased out. There is a strong incentive toward source reduction and recycling, which creates a need for separations technology [1].

All of the above separation needs are oriented primarily toward removal and isolation of hazardous material from effluent or waste streams. Pollutants are frequently present in only trace quantities, such that highly resolving separation systems will be required for detection and removal. The problem of removing pollutants from extremely dilute solutions is becoming more important as allowable release levels for pollutants are lowered. For example, proposed standards for the release of arsenic prescribe levels at or below the current limit of detection. Another example is pollution of water with trace quantities of dioxin. In research being carried out at Dow Chemical USA, concentrations of adsorbed dioxin at the part-per-quadrillion (10^{15}) level have been successfully removed from aqueous effluents. That technology has now been scaled up, such that dioxin removals to less than ten parts per quadrillion are being achieved on a continuous basis on the 20 million gallon per day wastewater effluent stream from Dow's Midland, Michigan, manufacturing facility.

1.5   Environmental separations

Based upon sources of pollution and the nature of polluted sites (air, land, or water), environmental separations can be categorized as follows.

1 Clean up of existing pollution problems

Examples:
- surface water contamination (organics, metals, etc.)
- groundwater contamination (organics, metals, etc.)
- airborne pollutants (SO_x, NO_x, CO, etc.)

- soil clean-up (solvent contamination, heavy metals, etc.)
- continuing discharges to the environment
    automobiles
    industries (chemical, nuclear, electronics, engineering, etc.).

2 Pollution prevention

Examples:

- chemically benign processing
    hybrid processing
    use of water instead of hydrocarbon/fluorocarbon solvents
    alternative chemical synthesis routes
- use of separation step(s)
    reduction in downstream processing steps
    eliminate solvent use (membranes instead of extraction, for example)
    eliminate purge streams (internally remove contaminants so purge stream is not
        needed)
    recovery and recycle instead of discharge (organics, water).

Figure 1.1 portrays a hierarchy for pollution prevention [5]. It is apparent that the difficulty of implementation decreases from top to bottom. Note that, the first four approaches on the hierarchy involve chemical separations (mass transfer operations).

The Chemical Manufacturers Association has published a strategy [6] for addressing pollution minimization or elimination in chemical processing facilities very similar to Figure 1.1. They suggest, in priority order:

1 Source reduction. Process changes to eliminate the problem.
    These process changes can include:
    - Reducing by-product formation through changes in processing and/or catalyst usage. This step can include changes in raw materials used.
    - Better process control to minimize processing variations which lead to additional discharges.
    - New processing flowsheets to minimize unwanted product generation and/or release.
2 Recycle. If source reduction is not feasible, then recycle
    - within the process
    - within the plant
    - off-site.
3 Treatment. Post-process waste treatment prior to discharge to minimize the environmental impact.

A recent article [7] describes more than 50 pollution prevention strategies that do not require large investment costs.

The use of chemical separations is already very important in many industries. These include biotechnology, metals recovery and purification, fuels, chemical processing plants and feedstocks, municipal sewage treatment, and microelectronics. For these and other industries, the efficiency of the separation steps is often the critical factor in the final cost of the product.

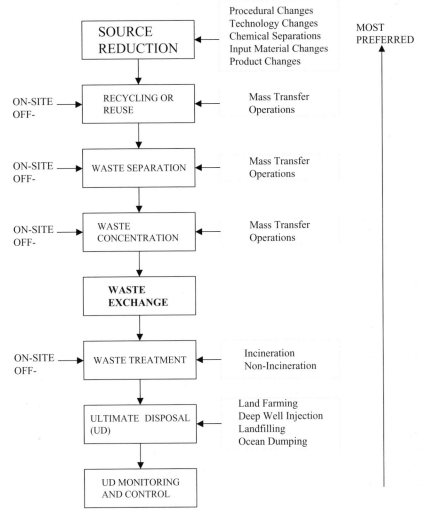

Figure 1.1 Pollution Prevention Hierarchy [5]. (Copyright © 1993, John Wiley & Sons, Inc.) This material is used by permission of John Wiley & Sons, Inc.

The separation cost is often related directly to the degree of dilution for the component of interest in the initial mixture. This cost includes the fact that most separations use 50 times the minimum energy requirement based on the ideal thermodynamic requirements. To put the energy consumption in perspective, the chemical and petroleum refining industries in the US consume approximately 2.9 million barrels per day of crude oil in feedstock conversion [1]. One method to visualize this cost factor is with the Sherwood plot shown in Figure 1.2.

This log–log plot shows that there is a reasonable correlation between the initial concentration of a solute in a mixture and its final price. For environmental applications, this correlation would translate to the cost of removal and/or recovery of a pollutant based on its initial concentration.

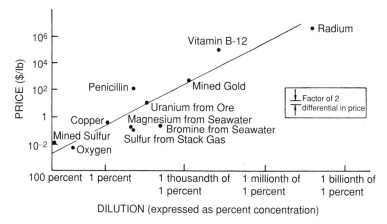

Figure 1.2 Sherwood plot [1]. Reproduced with permission of National Academy Press.

1.6   Historic perspective of environmental pollution

Rainwater is acidic due to atmospheric CO_2, SO_2 and nitrogen oxides; its pH is typically 5.6. Measurements of 4.6 are found in some regions of the US and values of 4.0 (and even 3.0) have been documented. Since pH is a log scale, these low pH values represent much stronger acids than occur naturally. The effects of these stronger acids on plants, animals, and materials have been well documented. Acid deposition can initially be dry. Gases and/or salts can be deposited. They can cause damage "as is," such as uptake by plants, or when hydrated [8]. In addition to contributing to acid rain, CO_2 also acts as a "greenhouse gas" and contributes to global warming.

The issue of chemical emissions and their effect on the environment is not limited to recent history. As shown below, acid rain was first documented in the 1600s. The chronology below lists some important events in the identification, monitoring, and steps to reduce emissions for acid rain and global warming [9].

1661–2   English investigators John Evelyn and John Graunt publish separate studies speculating on the adverse influence of industrial emissions on the health of plants and people. They mention the problem of transboundary exchange of pollutants between England and France. They also recommend remedial measures such as locating industry outside of town and using taller chimneys to spread "smoke" into "distant parts."

1734   Swedish scientist C. V. Linné describes a 500-year-old smelter at Falun, Sweden: "... we felt a strong smell of sulfur ... rising to the west of the city ... a poisonous, pungent sulphur smoke, poisoning the air wide around ... corroding the earth so that no herbs can grow around it."

1872   English scientist Robert Angus Smith coins the term "acid-rain" in a book called *Air and Rain: The Beginnings of a Chemical Climatology*. Smith is the first to note acid-rain damage to plants and materials. He proposes detailed procedures for the collection and chemical analysis of precipitation.

1911   English scientists C. Crowther and H.G. Ruston demonstrate that acidity of precipitation decreases the further one moves from the center of Leeds, England. They associate these levels of acidity with coal combustion at factories in Leeds.

1923   American scientists W.H. MacIntyre and I.B. Young conduct the first detailed study of precipitation chemistry in the United States. The focus of their work is the importance of airborne nutrients to crop growth.

1948   Swedish scientist Hans Egner, working in the same vein of agricultural science as MacIntyre and Young, set up the first large-scale precipitation chemistry network in Europe. Acidity of precipitation is one of the parameters tested.

1954   Swedish scientists Carl Gustav Rossby and Erik Eriksson help to expand Egner's regional network into the continent-wide *European Air Chemistry Network*. Their pioneering work in atmospheric chemistry generates new insights into the long-distance dispersal of air pollutants.

1972   Two Canadian scientists, R.J. Beamish and H.H. Harvey, report declines in fish populations due to acidification of Canadian lake waters.

1975   Scientists gather at Ohio State University for the *First International Symposium on Acid Precipitation and the Forest Ecosystem*.

1977   The UN Economic Commission for Europe (ECE) sets up a *Cooperative Programme for Monitoring and Evaluating the Long-Range Transmission of Air Pollutants in Europe*.

1979   The UN's World Health Organization (WHO) establishes acceptable ambient levels for SO_2 and NO_x. Thirty-one industrialized nations sign the *Convention on Long-Range Transboundary Air Pollution* under the aegis of the ECE.

1980   The US Congress passes an *Acid Deposition Act* providing for a 10-year acid-rain research program under the direction of the *National Acid Precipitation Assessment Program*.

1980   The United States and Canada sign a Memorandum of Intent to develop a bilateral agreement on transboundary air pollution, including "the already serious problem of acid rain."

1985   The ECE sets 1993 as the target date to reduce SO_2 emissions or their transboundary fluxes by at least 30% from 1980 levels.

1986   On January 8, the Canadian and US Special Envoys on Acid Rain present a joint report to their respective governments calling for a $5 billion control technology demonstration program.

1986   In March, US President Ronald Reagan and Prime Minister Brian Mulroney of Canada endorse the *Report of the Special Envoys* and agree to continue to work together to solve the acid-rain problem.

1995   An Intergovernmental Panel on Climate Change, representing over 2,000 scientists from over 50 countries, concludes that "the balance of evidence suggests there is a discernable human influence on global climate." They also list some striking projections by 2100 if the present trends continue:

- greenhouse gases could exceed 700 ppm levels not seen for 50 million years
- average atmospheric temperature will rise by 2 to 6.5 °F (1 to 3.5 °C), exceeding the rate of change for the last 10,000 years
- sea levels could rise between 6 to 37 inches (0.15 to 1 m).

1997   *Kyoto Protocol* agreement reached. This agreement is the first global approach to controlling greenhouse gas emissions.

Separations technology is already making an important contribution to ameliorating the acid-rain problem. Wet-scrubbing processes are the most widely used systems for removal of sulfur and nitrogen compounds from effluent stack gases. The limits of cost for wet-scrubbing techniques are such that they are not used to remove more than 75 percent of the sulfur-oxide compounds present and are currently of only limited effectiveness for removal of nitrogen oxides. Such systems also produce large quantities of sludge that present a solids disposal problem. New reagent systems that can be used in a more effective recycling mode are needed, and would be particularly useful if they could simultaneously remove both sulfur and nitrogen compounds in forms from which they could be converted into useful products. In any case, effective approaches must be brought into use to remove the nitrogen compounds.

1.7   The sulfur problem: where separations can help

Our principal sources of energy – fossil fuels – are all contaminated to some extent with sulfur compounds. When these fuels are burned, the sulfur compounds are burned to sulfur oxides, which are emitted to the atmosphere in the flue gas. In the atmosphere, these oxides are converted into the sulfur acids that are a principal cause of acid rain.

Separations technology plays a critical role in limiting sulfur-oxide pollution from sulfur-bearing fossil fuels. This technology is sufficiently advanced that there are no inherent technological limits to removing more than 95 percent of the sulfur present in natural gas, crude oil, and coal – many processes exist for accomplishing this before, during, or after combustion. The principal barriers to nearly complete sulfur removal are cost and practicality.

Natural Gas.   The principal sulfur contaminant of natural gas is another gas – hydrogen sulfide. Because it is extremely toxic, civil authorities have long forbidden significant levels of this compound in natural-gas pipelines. Hydrogen sulfide is removed from natural gas by a variety of commercial processes including reaction with aqueous solutions of oxidants, absorption into aqueous solutions of bases, distillation, and selective permeation through membranes. The end product of these processes is elemental sulfur, which can be sold and, in some cases, is worth more than the co-produced natural gas. In 1984, about 24,000 tons (24 million kilograms) of sulfur was produced from natural-gas wells in the United States.

Petroleum.   Sulfur can also be recovered from crude oil with technology that relies on the reaction of hydrogen with sulfur-containing compounds in crude oil (hydrodesulfurization) and permits modern refiners to turn 3 percent sulfur crudes into liquid product with no more than 0.5 percent sulfur. About 26,000 tons of saleable by-product sulfur was produced from crude oil in 1983.

Coal.   Coal can be partially desulfurized before combustion. Washing and magnetic separation are effective in reducing the content of iron sulfide, the principal inorganic sulfur contaminant, by up to 50 percent or somewhat higher. However, there are also organic sulfur compounds in coal, and a feasible means of removing them has not yet been found. Accordingly, combustion of coal produces a flue gas that contains significant amounts of sulfur oxides, which must be removed from the gas if sulfur pollution is to be minimized.

Flue-gas scrubbers are proven but expensive separation devices for removing sulfur from combustion gases. The new dry-scrubber technology removes about 90 percent of the sulfur in a flue gas by contact with a lime slurry in a specially designed combination spray dryer and reactor. The reaction product is a dry calcium sulfate–sulfite mix that is environmentally benign. Larger users favor the wet-scrubber technology, which is capable of removing up to 90 percent of the sulfur with a lime slurry in a contactor column.

Separations technology has made a substantial contribution to reducing the sulfur-pollution problem associated with the burning of fossil fuels. The principal barrier to further alleviation of this problem is economic and will respond to improved technology gained through further research and development [1].

1.8   Remember

- Environmental separations can apply to the clean-up of existing problems as well as pollution prevention.
- The cost of separations is directly related to the degree of dilution in the feed stream.
- The three primary functions of separation processes are purification, concentration, and fractionation.
- Separations use thermodynamic equilibrium- and/or mass transfer (rate-) based analysis.

1.9   Questions

1.1   Give three examples of pollution sources for (a) water; (b) air; (c) soil.
1.2   Using the Sherwood plot, what is the price differential for a product contained in a 1% and a 0.001% feed stream?
1.3   Give two examples of a separation process that can be analyzed based on (a) thermodynamic equilibrium; (b) mass transfer (rate).

1.4 Based on the Pollution Prevention Hierarchy, what is the best approach to minimize pollution?

1.5 Give two examples of (a) pollution prevention; (b) clean-up of existing pollution problems.

1.6 Describe, in your own words, why you are interested in environmental applications.

1.7 What do you think is the future of separations for environmental applications?

1.8 List and describe three external stimuli that can cause movement of matter.

1.9 Which is more difficult to control and remediate, point-source or non-point-source pollution? Why? What about mobile versus stationary sources?

1.10 Electric cars are recharged at night by electricity from local power plants. If local power plants generate electricity by burning fossil fuels, why are electric cars more environmentally sound than conventional gas (petrol)-burning cars?

1.11 Explaining the pollution cost of energy production and use from an environmental point of view, why is nuclear generated power better than conventional power plants? What waste products are generated by nuclear power?

1.12 Fifty years ago, pollution control was governed by the quote, "the solution to pollution is dilution." Why is this no longer an acceptable practice?

1.13 Why is source reduction the best practice for pollution control?

1.14 Discuss the effects of population growth on the environment.

1.15 The adsorption step in a process sequence can be used to remove a contaminant from a fluid stream by contact with a stationary solid adsorbent bed. Is this an example of purification, concentration, or fractionation, and why?

1.16 Is it possible for a separation process used to control one form of pollution to create another? Explain.

2

Separations as unit operations

The goal is to pick the best solution.

– H. S. FOGLER

The information in the previous chapter provides an important introduction to the environmental applications of chemical separations technology. This chapter will be devoted to an introductory description of the concept and analysis of a unit operation as applied to separation processes. Subsequent chapters will present some necessary fundamentals of separations analysis and discuss specific separation methods.

2.1 Objectives

1 Define the concept of a unit operation and state the design significance.
2 Describe the two basic mechanisms for separations.
3 Discuss factors important in selecting an exploitable property difference.
4 Give examples of equilibrium and rate properties that are used as the basis for separation.
5 Give examples of mass- and energy-separating agents.
6 List the two ways that a separating agent is used to obtain a different compound distribution between two phases.
7 List the four ways that separating agents generate selectivity.
8 Discuss the applications of reversible chemical complexation to separations.
9 Define cocurrent and countercurrent operation.
10 List factors important to the selection of a particular separation process for a given application.
11 List several reasons for implementing a unified view of separations technologies.

2.2 Unit operations

Initially, it is useful to introduce the concept of a unit operation and explain how it relates to chemical separations. Figure 2.1 shows a generic unit operation in which a feed stream is separated into two exit streams with different compositions by means of a separating agent. Multiple feed streams into a process and multiple exit streams are also possible. The separating agent can be either a mass or an energy modification. Separating agents exploit a physical property difference to facilitate the separation. The mechanism for separation uses this physical property difference to provide the separation. Later in this chapter, the concept of a mass-separating agent will be discussed in detail. The separation that occurs will depend upon process conditions such as feed composition, phase, temperature, pressure, flowrates, the separating agent, and the separation method used.

It is important to note that the fate of each stream is important with respect to environmental impact. This includes the separating agent. The use of energy as a separating agent has a direct impact through energy consumption but also generates pollution due to energy production. The environmental impact of a mass-separating agent involves the ultimate fate of the mass. How is it ultimately disposed? Does it enter the product streams?

A unit operation is any single step in an overall process that can be isolated and that also tends to appear frequently in other processes. For example, a car's carburetor is a single unit operation of the engine, just as the heart is a unit operation of the human body. The concept of a unit operation is based on the idea that general analysis will be the same for all systems because individual operations have common techniques and are based on the same scientific principles. In separations, a unit operation is any process that uses the same separation mechanism. For example, adsorption is a technique in which a solid sorbent material removes specific components, called solutes, from either gas- or liquid-feed streams because the solute has a higher affinity for the solid sorbent than it does for the fluid. The mathematical characterization of any adsorption column is the same regardless

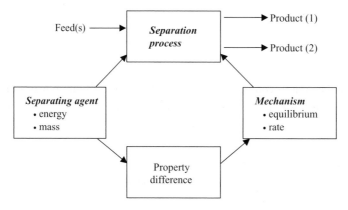

Figure 2.1 Generic unit operation for a separation process.

of which solutes are being removed from a given fluid by a given sorbent or the amount of fluid processed in a given time. Hence, the design and analysis of a particular separation method will be the same regardless of the species and quantities to be separated. Once it is understood how to evaluate an adsorption column to separate a binary component feed stream, the same principles can be applied to any binary mixture. An important aspect of design and analysis is the scale (or size) of a process. Those in which separations technology is required span several orders of magnitude in terms of their throughput. For example, industrial separation of radioisotopes occurs at a production rate on the order of 10^{-6} to 10^{-3} kg/hr, while coal cleaning plants operate at a production rate greater than 10^6 kg/hr. If new design criteria had to be developed for a given separation technique each time that the scale of the process was changed, the analysis would be of very limited value and one would have to write a book for each separation technology to cover all the potential process sizes. The concept of a unit operation, therefore, allows us to apply the same design criteria and analysis for a given separation technology, irrespective of the size. This very important element in evaluation allows one to scale-up or scale-down a process based upon results obtained on a different-sized piece of equipment. This is the basis for conducting tests on bench- or pilot-scale equipment and using the results for the design of the full-scale process. In addition, bench or pilot-plant test results can be used to determine the effect of a single separation step or other unit operation on an overall process. The configuration and flow patterns of any single step can affect the entire process and are usually determined experimentally.

2.3 Separation mechanisms

Separation processes rely on various mechanisms, implemented via a unit operation, to perform the separation. The mechanism is chosen to exploit some property difference between the components. They fall into two basic categories: the partitioning of the feed stream between phases; and the relative motion of various chemical species within a single phase. These two categories are often referred to as equilibrium and mass transfer rate processes, respectively. Separation processes can often be analyzed with either equilibrium or mass transfer models. However, one of these two mechanisms will be the limiting, or controlling, factor in the separation and is, therefore, the design mechanism.

For a separation to occur, there must be a difference in either a chemical or physical property between the various components of the feed stream. This difference is the driving force basis for the separation. Some examples of exploitable properties are listed in Table 2.1. Separation processes generally use one of these differences as their primary mechanism.

The following factors are important considerations in the choice of a property difference.
(a) Prior experience. The reliability and "comfort" factor go up if there has been prior <u>positive</u> experience in the use of a certain property difference for certain applications.
(b) The property itself. How simple will it be to implement?

Table 2.1 *Exploitable properties used in separation processes.*

Equilibrium properties	Rate properties
Vapor pressure	Diffusivity
Partitioning between phases	Ionic mobility
solubility of a gas in a liquid	Molecular size and shape
sorption of a solute in a fluid onto a sorbent	
Chemical reaction equilibrium	
Electric charge	
Phase change	
solid/liquid	
liquid/gas	

Table 2.2 *Examples of separating mechanisms and separating agents.*

Separation process	Separation mechanism	Separating agent
Distillation	Vapor pressure	Heat
Extraction	Partitioning between phases	Immiscible liquid
Adsorption	Partitioning between phases	Solid sorbent
Absorption	Partitioning between phases	Non-volatile liquid
Filtration	Molecular size and shape	Membrane
Ion exchange	Chemical reaction equilibrium	Solid ion-exchanger
Gas separation	Diffusivity and phase partitioning	Membrane
Electrodialysis	Electric charge and ionic mobility	Charged membrane/electric field

(c) The magnitude of the property difference. Obviously, the larger the difference between the components to be separated, the easier the separation. How large is good enough? This answer is based on experience and the value of the components to be separated. Preliminary calculations for various separation processes (using various property differences) can be very useful.

(d) Chemical behavior under process conditions. Will the process fluids chemically attack the separation equipment (corrosion, morphological changes such as swelling, etc.) and/or react themselves (polymerize, oxidize, etc.)? This is a very important consideration as it affects the lifetime and reliability of the process.

(e) Quantities and phases which need to be processed. This is an economic consideration as the cost of implementing various property differences (energy input as heat, for example) is a function of the scale of the process. The phases (gas vs liquid, for example) also affect the equipment size and material handling considerations.

(f) Separation criteria. What are the concentrations of the various components in the feed stream? What purity and recovery are needed? How many components in the feed need to be separated?

Table 2.2 is a listing of several common separation processes, their primary separation mechanisms, and the separating agent used. The separating agent concept will be explained in some detail in a later section of this chapter.

2.4   Equilibrium-based processes

In an equilibrium-based process, two phases (vapor, liquid, or solid) are brought into contact with each other, mixed thoroughly, then separated with a redistribution of the components between phases. Often multiple contacts are made in a series of cascading steps in which the two phases flow countercurrent to each other. At each contact the phases are allowed to approach thermodynamic equilibrium. Once equilibrium is reached, there can be no more separation without a change in the operating parameters of the system that affect the equilibrium relationship. The next stage in the cascade, therefore, has at least one process change that alters the equilibrium relationship to establish a new equilibrium relationship. The cascade should be designed such that conditions are altered at each stage to move closer toward the desired separation. For example, distillation (Chapter 4) is a fractionating separation in which a binary (or multiple) feed stream is separated into two (or more) product streams based upon their differences in boiling point. One type of distillation column has a series of cascading contact trays such that the temperature increases from the top tray, which is just above the boiling point of the lower-boiling-point component, to the bottom tray, which is just below the boiling point of the higher-boiling-point component. Thus, the lower-boiling-point component is enriched in the gas phase, while the higher-boiling-point component is enriched in the liquid phase. Each tray from the top to the bottom of the column, then, operates at a higher temperature such that a new equilibrium is established down the length of the column. As the temperature increases down the column, the lower-boiling-point species tends to vaporize more and move up the column as a gas stream, while the higher-boiling-point component continues down the column as a liquid. The final result is a vapor stream leaving the top of the column which is almost pure in the lower-boiling-point species and a liquid stream exiting the bottom of the column that is almost pure in the higher-boiling-point component.

For phase partitioning (equilibrium), the variable of interest is the solute concentration in the first phase that would be in equilibrium with the solute concentration in a second phase. For example, in the distillation example above, each component is partitioned between the vapor and liquid phases. The mathematical description of the equilibrium relationship is usually given as the concentration in one phase as a function of the concentration in the second phase as well as other parameters. Some examples are the Henry's Law relation for the mole fraction of a solute in a liquid as a function of the mole fraction of the solute in the gas phase which contacts the liquid:

$$y = mx, \tag{2.1}$$

where y = mole fraction in the gas phase

  x = mole fraction in the liquid phase

  m = Henry's Law constant.

A second example would be the Langmuir isotherm characterizing adsorption (Chapter 7), which relates the equilibrium amount of a solute sorbed onto a solid to the concentration in the fluid phase in contact with the solid:

$$X = \frac{K_s bC}{1 + bC},$$ (2.2)

where X = amount of solute sorbed per weight of sorbate

  C = solute concentration in fluid phase

  K_s, b = constants.

An important factor in the use of phase partitioning for separations is the degree of change in composition between the two phases. In the limit where the composition in each phase is identical, separation by this mechanism is futile. For vapor–liquid equilibrium, the condition is called an azeotrope. Irrespective of the phases, this condition corresponds to a partition coefficient of unity.

Some data and model equations will be provided in the appropriate chapters. Appendix E provides additional information and references for data and calculation methods.

2.5 Rate-based processes

Rate-based processes are those in which one component of a feed stream is transferred from the feed phase into a second phase due to a gradient in a physical property. Gradients in pressure or concentration are the most common. Other gradients include temperature, electric fields, and gravity. The limiting step upon which design is based is the rate of transfer of the particular component from the feed material to the second phase. For relative motion (rate) of the various chemical species, the mathematical description relates the rate of transfer of a particular component across a boundary due to a driving force. One example is Fick's Law that relates the flux of a component (N_A) across a layer (fluid or solid) to the concentration gradient within the layer:

$$N_A = -D_A \frac{dC_A}{dx},$$ (2.3)

where D_A = diffusion coefficient of A in the medium (physical property found in many handbooks)

dC_A/dx = concentration gradient of A in the direction of interest.

A second example is the use of a mass transfer coefficient to relate the flux across a fluid boundary layer (fluid region over which the solute concentration changes from the bulk phase value) to the concentration difference across the layer:

$$N_A = k(C_{A,1} - C_{A,2}), \tag{2.4}$$

where N_A = mass flux of A across the fluid layer

$\quad k$ = mass transfer coefficient

$\quad C_A$ = concentration of component A.

In choosing between these two models, one needs to consider the specific process. The use of mass transfer coefficients represents a lumped, more global view of the many process parameters that contribute to the rate of transfer of a species from one phase to another, while diffusion coefficients are part of a more detailed model. The first gives a macroscopic view, while the latter gives a more microscopic view of a specific part of a process. For this reason, the second flux equation is a more engineering representation of a system. In addition, most separation processes involve complicated flow patterns, limiting the use of Fick's Law. A description of correlations to estimate values of k for various systems is contained in Appendix B.

2.6 Countercurrent operation

The analysis of equilibrium-stage operations is normally performed on the basis of countercurrent flow between two phases. Because most separation processes, whether described in terms of equilibrium or mass transfer rates, operate in this flow scheme, it is useful to compare countercurrent to cocurrent flow. Figure 2.2 illustrates cocurrent and countercurrent operation. Assuming mass transfer across a barrier between the two fluid phases, generic concentration profiles can be drawn for each case (Figure 2.3).

A few points become obvious. First, in each case, the concentration difference across the barrier changes with axial position x. So, the flux (or rate) will change with position. Second, for cocurrent flow, the concentration difference (driving force for mass transfer) becomes very small as the flow moves axially away from the entrance ($x = 0$). So, the separation becomes less efficient as the barrier becomes longer in the axial direction. For countercurrent operation, the driving force is maintained at a larger value along the

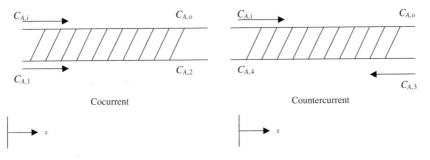

Figure 2.2 Cocurrent vs countercurrent operation.

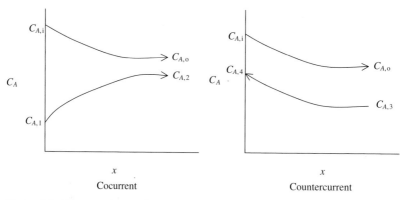

Figure 2.3 Concentration profiles.

barrier and $C_{A,4}$ can be larger than $C_{A,o}$. Therefore, countercurrent operation is usually the preferred method.

To account for the variation in driving force, a log–mean driving force is used instead of a linear one:

$$\Delta C_{lm} = \frac{\Delta C_1 - \Delta C_2}{\ln\left(\Delta C_1 / \Delta C_2\right)}. \tag{2.5}$$

This equation is derived in Middleman [1].

For countercurrent flow:

$$\Delta C_1 = C_{A,i} - C_{A,4}$$
$$\Delta C_2 = C_{A,o} - C_{A,3}. \tag{2.6}$$

For cocurrent flow:

$$\Delta C_1 = C_{A,i} - C_{A,1}$$
$$\Delta C_2 = C_{A,o} - C_{A,2}. \tag{2.7}$$

ΔC_1 and ΔC_2 are the concentration differences at each end of the barrier. This analysis will be used in the chapter on membranes (Chapter 9).

2.7 Productivity and selectivity

In the evaluation of a separation process, there are two primary considerations: productivity and selectivity. The productivity, or throughput, of a process is the measure of the amount of material which can be treated by this process in a given amount of time. This quantity is usually specified by the feed flowrate to the process and/or the amount of a product stream. The selectivity of the process is the measure of the effectiveness of the process to separate the feed mixture. Selectivity is usually given by a separation factor (α_{ij}), which is a ratio of compositions in the product streams for an equilibrium process or rates of mass transfer for a rate-based process:

$$\alpha_{ij} = \frac{x_{i,1}/x_{j,1}}{x_{i,2}/x_{j,2}}, \tag{2.8}$$

where $x_{i,1}$ = fraction of component i in stream 1

$\quad x_{j,1}$ = fraction of component j in stream 1

$\quad x_{i,2}$ = fraction of component i in stream 2

$\quad x_{j,2}$ = fraction of component j in stream 2.

Various terms will be used to represent these quantities depending on the process. A target for the composition of one or more product streams usually dictates the separation requirement for a particular process.

There are two types of separation factors commonly used; the ideal and the actual separation factors. The ideal separation factor is based on the equilibrium concentrations or transport rates due to the fundamental physical and/or chemical phenomena that dictate the separation. This is the separation factor that would be obtained without regard to the effects of the configuration, flow characteristics, or efficiency of the separation device. This value can be calculated from basic thermodynamic or transport data, if available, or obtained from small-scale laboratory experiments. For an equilibrium-based separation, the ideal separation factor would be calculated based on composition values for complete equilibrium between phases. For a rate-based separation, this factor is calculated as the ratio of transport coefficients, such as diffusion coefficients, without accounting for competing or interactive effects. Each component is assumed to move independently through the separation device.

2.7.1 Equilibrium-based process

The separation factor, α_{ij}, is the ratio of the concentration of components i and j (mole fractions, for example) in product stream (1) divided by the ratio in product stream (2). For example, in distillation α_{ij} is defined in terms of vapor and liquid mole fractions (Figure 2.4):

$$\alpha_{ij} = \frac{x_{i,1}/x_{j,1}}{x_{i,2}/x_{j,2}} = \frac{y_a/y_b}{x_a/x_b}, \tag{2.9}$$

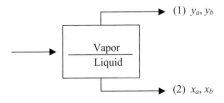

Figure 2.4 Distillation product streams.

where y_a, y_b = the mole fractions of components a and b in the overhead stream

x_a, x_b = the mole fractions of a and b in the bottoms stream. [Note: not the feed conditions.]

For example: $x_a = x_b = 0.5$, $y_a = 0.6$, $y_b = 0.4$

$$\alpha_{ab} = \frac{0.6/0.4}{0.5/0.5} = \frac{1.5}{1.0} = 1.5.$$

<u>Ideal</u>: Based on vapor pressure of each component.

<u>Actual</u>: Account for non-idealities in solution (fugacity vs pressure, for example).

2.7.2 Rate-based process

Component flowrates can be used for rate processes, as shown in Figure 2.5.

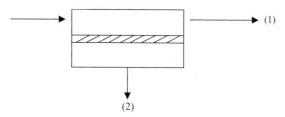

Figure 2.5 Membrane separation product streams.

Membrane separation

In computing the separation factor, one must use appropriate physical parameters, such as operating conditions and equipment size (membrane area in this case) to relate the flux to a driving force. The compositions of streams (1) and (2) may be used; however, it is better to use the ratio of permeabilities, transport coefficients, or other measures of the inherent separating ability of the device. One can think of α as a flux ratio scaled for a unit driving force:

$$\alpha_{ab} = \frac{(\text{flux/driving force})_a}{(\text{flux/driving force})_b} = \frac{Q_a}{Q_b}. \tag{2.10}$$

<u>Ideal</u>: Based on single-component measurements. Normally, does not account for configuration or flow characteristics of the separation device.

<u>Actual</u>: Would include any competitive effects, interactive effects and effects of device.

The separation factor is usually given as a value of one or greater. The selectivity of the separation is improved as the value of this ratio is increased.

When one determines the separation for an actual feed mixture in a separation device, an actual separation factor is obtained. This value is usually obtained by measurements on the device and is usually less than the ideal value ($\alpha_{\text{actual}} < \alpha_{\text{ideal}}$).

2.8 Separating agents

Many separation processes are based on the formation of an additional phase which has a different composition from the feed stream(s). One possible way of forming another phase is the addition of energy (energy-separating agent) to convert a liquid stream to a vapor stream. Distillation exploits this idea to separate mixtures of liquids that boil at different temperatures. Crystallization processes use energy to separate liquid mixtures with components that solidify at different temperatures. The temperature is lowered until the species with the higher solidification temperature crystallizes out of solution. Evaporation and drying are other processes in which energy addition promotes the separation by formation of a new phase.

Another large class of separations makes use of a change in solute distribution between two phases in the presence of mass not originally present in the feed stream. This mass-separating agent, MSA, is added as another process input to cause a change in solute distribution. The MSA can alter the original phase equilibrium or facilitate the formation of a second phase with a concentration of components different from that in the original phase. One of the components of the original feed solution must have higher affinity for the MSA than for the original solution. This solute will then preferentially transfer from the original feed solution to the MSA phase. Once the MSA has been used to facilitate a separation, it must normally be removed from the products and recovered for recycle in the process. Hence, use of an MSA requires two separation steps: one to remove a solute from a feed stream; and a second to recover the solute from the MSA.

General flowsheets for two basic processes using MSAs are shown in Figure 2.6. In each case, the solute in the feed fluid has a high affinity for the MSA. In Figure 2.6(a), the MSA is recirculated between two beds, one in which the solute is being sorbed and one in which the MSA is being regenerated. In this scheme, the MSA is a moving portion of the system and each vessel serves only one purpose; either the separation or the regeneration of the MSA. In Figure 2.6(b), the MSA is fixed in each bed. Both beds are capable of operating in either the sorption or regeneration mode. While one is in the sorption mode, the other is regenerating the MSA. When the first is saturated, they switch roles. The first scenario is common in cases in which the MSA is easily transported, as in the case of a liquid or gas solvent. The second case is more common for solid MSAs that are not easily transported.

As already stated, MSAs can consist of solids, liquids or gases. Figure 2.7 shows various combinations of feed phase and MSA phase with examples of various separation processes involved. In almost all cases, the use of an MSA involves the two steps shown above.

A separation involving an energy-separating agent (ESA) can involve input and removal steps, such as in distillation, where there is a reboiler for energy input and a condenser for energy removal. In other cases, such as evaporation, the vapor can be discharged without

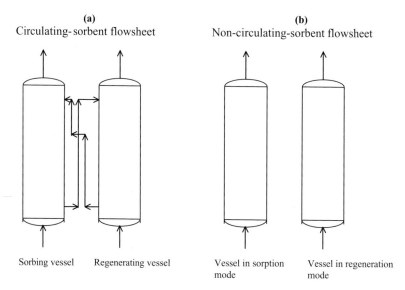

Figure 2.6 "Generic" flowsheets for MSA-based processes [2]. Reproduced with permission of the American Institute of Chemical Engineers. Copyright © 1987 AIChE. All rights reserved.

| | **Mass-Separating Agent** | | |
	Solid	Liquid	Gas
Feed			
Solid		Leaching	Steam Stripping
Liquid	Ion Exchange	Solvent Extraction	Stripping
Gas	Adsorption	Absorption	

Figure 2.7 Examples of separation processes using mass-separating agents.

the need for a heat removal step. An energy-separating agent has the advantage that no additional material is introduced into the system.

While heat is the most common energy-separating agent used, Table 2.3 provides examples of separation processes using gravitational, electric and magnetic fields.

Some general statements regarding the use of separating agents can now be made.

• Separation processes use mass- and/or energy-separating agents to perform the separation. Mass-separating agents can be a solid, liquid or gas. Heat is the most common

Table 2.3 *Some examples of separation processes using other than conventional fields to facilitate separation [2]. Reproduced with permission of the American Institute of Chemical Engineers. Copyright 1987 AIChE. All rights reserved.*

Process	Uses
Centrifugation	Separation of finely divided solids from liquids. Separation of liquid–liquid dispersions. Gaseous separation of isotopes.
Electrophoresis	Separation of charged macromolecules. Separation of mixtures of simple cations or anions.
Electrodialysis	Desalting of brackish water. Concentration of seawater.
Electrostatic separation	Removal of particulates from air and gas streams. Concentration and separation of various ores.
Electrosorption	Desalination.
High-gradient magnetic separation	Elective separation of weakly magnetic materials. Separation of inorganic-sulfur-containing particles from coal.
Field flow fractionation	Separation of particles by size in a shear field.

energy-separating agent, used in distillation. External fields, such as magnetic and electric, are sometimes used as energy-separating agents.

- **A different component distribution between two phases is obtained.** This distribution change can be accomplished in two ways:

  1 Alteration of the original phase equilibrium. Two phases are originally present and the role of the separating agent is to change the composition in each phase relative to the initial values.

  Examples:

  Mass-separating agent: one approach would be the addition of a selective complexing agent to a liquid phase to increase the solubility of a solute. An example would be the addition of a selective chelating agent to an organic phase which can complex with a metal ion in an aqueous phase when the two phases are contacted.

  Energy-separating agent: heat input to change the temperature of a gas/liquid system. The most common example is distillation.

  2 Generation of a second phase with a different component distribution.

  Examples:

  Mass-separating agent: solid sorbent which is selective for a solute in a feed stream. An example would be the removal of VOCs from an air stream using activated carbon.

  Energy-separating agent: heat input to a liquid phase to change the solute solubility in that phase. An example would be evaporation to remove water from a waste stream and concentrate it prior to disposal.

- Separating agents employ four methods to generate selectivity:

  1 Modification of phase equilibrium.

    Mass-separating agent: ion-exchange resin to selectively partition ions into the resin.

    Energy-separating agent: heat removal to precipitate salts from an aqueous stream.

  2 Geometry differences.

    Mass-separating agent: membrane which filters suspended solids based on size.

    Energy-separating agent: gravity settling to separate particles by size.

  3 Kinetics (rate of exchange) between phases.

    Mass-separating agent: use of amines in gas sorption to change the rate of acid gas (CO_2, H_2S) uptake into the liquid phase.

    Energy-separating agent: heat input into the amine solution to accelerate the rate of gas desorption which regenerates the solution.

  4 Rate of mass transfer within a phase.

    Mass-separating agent: intra-particle diffusion in a porous sorbent.

    Energy-separating agent: application of an electric field across a liquid phase to accelerate the charged particles relative to neutrals.

Mass-separating agents are generally characterized by their capacity to incorporate the desired solute (sometimes called loading) and their ability to discriminate between solutes (selectivity). Energy-separating agents are usually described by the amount required to achieve a certain throughput (productivity) and selectivity for a given process. These values relate directly to the equipment size needed for a given separation.

2.9 Reversible chemical complexation

One important mechanism by which a mass-separating agent can operate is reversible chemical complexation. A complexing agent is used in the contacting phase as the MSA. It can reversibly and selectively form a chemical complex with a solute in the feed stream. The reversibility is important so that the solute can be recovered and the complexing MSA can be recycled for reuse. In this process, a solute in the feed phase would partition into the contacting phase and form a chemical complex with a complexing agent in a high-affinity form. The form of the complexing agent would be altered, by a change in temperature for example, to a low-affinity form, such that the complex would dissociate and the solute would be released into a receiving stream. The simplest complexation reaction that can illustrate this process is 1:1 binding between the solute and the complexing agent in its high-affinity state followed by dissociation of the complex in the low-affinity state of the complexing agent.

$$A + B_h \leftrightarrow AB_h, \tag{2.11}$$

where A = solute

B_h = complexing agent

K_h = equilibrium constant for reaction

k_h = forward reaction rate constant for reaction (2.11).

$$A + B_\ell \leftrightarrow AB_\ell, \tag{2.12}$$

where A = solute

B_ℓ = complexing agent in the low-affinity state

K_ℓ = equilibrium constant for low-affinity reaction

k_ℓ = forward reaction rate constant for reaction (2.12).

A change in solute concentration (partial pressure, for example), a change in temperature that changes the equilibrium constant of the reversible reaction, or a change in the oxidation state of the MSA, usually accomplishes the reversibility. For example, in pressure-swing adsorption, a change in system pressure alters the binding affinity of the solute to the solid sorbent (Le Chatelier's Principle). In the high-affinity state, the solute adsorbs onto the sorbent and is thus removed from the process fluid. Upon lowering the adsorbent bed pressure, the solute desorbs from the sorbent. This allows the solid sorbent to be regenerated for continued use without interrupting the overall system.

Reversible chemical complexation processes can be either equilibrium or mass transfer (rate) limited. For those in which equilibrium is the controlling, or design, mechanism, it is important that

$$K_h/K_\ell \gg 1. \tag{2.13}$$

For those in which rate of formation of the complex is the limiting factor,

$$k_h/k_\ell \gg 1. \tag{2.14}$$

The preferred usage is when the complexing agent interacts with the solute of interest and has little or no interaction with the other components of the feed stream. Separation processes based on reversible chemical complexation provide an enhancement in the solubility of the selected solute through the complexation reaction. This approach can provide high enhancement of capacities and selectivities for dilute solutes, especially when the solute feed concentration is below 10%.

There are several characteristics of a good complexing agent. First, for the reversibility of a complexation reaction to be easily accomplished, the bond energy for the association should be in the range of 7–70 kJ/mol. Second, there must be no side reactions. The complexation reaction must only take place with the solute of interest. Third, there can also be no irreversible or degradation reactions. Any reaction which decreases the amount of complexing agent available for the separation reduces the capacity and selectivity of the separation process. Thermal instability, oxidations, polymerizations, and reactions with the separation process materials of construction are all examples of potential problems. Fourth, a good complexing agent has no co-extraction of solvent from the feed phase. One example of an undesirable co-extraction is the case of liquid-phase metal extractions where water is extracted with the metal ion. This co-extracted water dilutes the metal concentration in the receiving phase. Fifth, it is important for a complexing agent to provide rapid kinetics.

Slow kinetics translates to additional time required for the extraction and/or release step. Sixth, and last, there can be no partitioning of the complexing agent into the feed or receiving phases. This effect would continuously decrease the amount of complexing agent available as the separation process is cycled.

2.10 Selection of a separation process

This section is normally included at the end of a textbook on separations; however, it is included here to give the reader some "food for thought" in deciding how they might use the material in the subsequent chapters. For additional perspectives on this topic, consult references [2–11]. The following should be taken as a heuristic or guide.

1 Assess the feasibility. What are the property differences that you plan to exploit for the separation(s)? Which processes use this property difference as their primary separating mechanism? What operating conditions are associated with the feed stream (flowrate, T, P, pH, reactive components, etc.)? Are these conditions "extreme" relative to normal operating conditions for a given separation process?

2 Determine the target separation criteria. What purity and recovery are needed for the various components in the feed stream? For a feasible separation process, what is the separation factor based on the property difference chosen? For an equilibrium-based process, this would be the separation factor for one stage.

   There are various molecular properties which are important in determining the value of the separation factor for various separation processes.

   (a) Molecular weight. Usually, the heavier a compound is, the lower the vapor pressure. Molecular weight is also related to molecular size, which affects diffusion rates and access to the interior of porous materials.

   (b) Molecular volume. This is a measure of density since there is an inverse relationship between density and volume. As will be seen in the analysis of various separation processes, density can be a significant variable.

   (c) Molecular shape. The molecular shape can certainly affect the access of certain molecules to pores and chemical binding sites. The shape also will affect how the molecules order in a liquid or solid phase.

   (d) Intermolecular forces. The strength of these forces can affect the vapor pressure and solubility in certain solvents. One property is the dipole moment which is a measure of the permanent charge separation within a molecule (polarity). Another property is the polarizability which is a measure of a second molecule's ability to induce a dipole in the molecule of interest. The dielectric constant is a physical property that can be used as a measure of both the dipole moment and polarizability.

   (e) Electrical charge. The ability of a molecule to move in response to an electric field is a function of the electrical charge.

   (f) Chemical complexation. Separations involving selective chemical reactions can impart higher selectivity (separation factors) than the use of a physical property

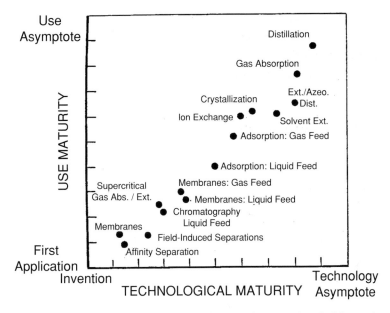

Figure 2.8 Separation technology reliability plot [2]. Reproduced with permission of the American Institute of Chemical Engineers. Copyright 1987 AIChE. All rights reserved.

difference alone. The relative ability of the various compounds in the feed stream to react will directly affect the separation selectivity.

3 Do the easiest separation first. This may seem obvious, but often the starting place is non-obvious when one is faced with a complex feed mixture to separate. What are some examples of "easy"?

   (a) <u>Simple</u>. If particles are present in a fluid phase, filtration can do the separation of the two phases. Separation of components in each phase can then be done in a subsequent step, if needed.

   (b) <u>High reliability</u>. Separation steps which have been used for long periods of time with positive results (high "comfort" factor) would be considered ahead of newer approaches (see Figure 2.8). Newer approaches are easier to implement further downstream where their impact on the entire process is less than as the first step.

   (c) <u>Remove component which has the highest mole fraction first</u>. If the separation factors for each component in the feed mixture are approximately equal, the component with the highest mole fraction is usually the easiest to remove first. An alternative is to remove the component with the highest volatility first if distillation is feasible.

   (d) <u>When more than one step is required for the separation sequence</u>.

      (i) Recover the mass-separating agent and/or dissolved products immediately after the process step involving a mass-separating agent.

      (ii) Do not use a mass-separating agent to remove or recover a mass-separating agent.

29

2.11 A unified view of separations

Having discussed the concept of unit operations in the context of separation, it is useful at this point to reiterate what is gained by this approach. King [12] lists several major gains in understanding, insight, capability, and efficiency that come from a unified view of separations. The first such gain recognized historically is that methods of analyzing the degrees of separation achieved in different separation processes are similar. Hence, a comprehensive knowledge of separations enables one to transfer the uses of separations to different scales of operation, ranging from analytical scale separations to large-process separations. In addition, the interactions of mass transfer and phase equilibria and their resultant effects are similar for related types of contacting equipment. The concepts used for stage efficiencies (fraction of equilibrium attained) are common, as well.

Developments of powerful computational algorithms have provided immense gains in computing capacity and the widespread use of personal computers has meant that much less attention has to be given to methods of calculating degrees of separation. This allows for much more emphasis on process selection, synthesis, and improvement. However, while many simulators exist to model separation processes, it is critical that the designer understands the fundamental principles prior to using them. A unified view makes it possible to identify and select among candidate separation processes for a given task on a knowledgeable basis. An understanding of patterns of stage-to-stage changes in composition in countercurrent separations and the causes for particular patterns is useful across the board for improvement of design and operating conditions to gain a greater degree of separation and/or lesser flows and equipment cost. In addition, an understanding of the factors governing energy consumption enables greater insight into reducing consumption and/or achieving an optimal combination of equipment and operating costs.

A general understanding of separations facilitates developing entirely new methods of separation. Insight into the capabilities of a variety of methods helps one to identify when the ability to separate will pose a major process limit. An understanding of solution and complexation chemistry makes it possible to identify and select among potential mass-separating agents for different applications and to transfer the use of particular agents and chemical functionalities among different types of separation processes.

2.12 Remember

- The concept of a unit operation allows us to develop design methodologies for various separation technologies that are the same for various sizes of equipment (scale-up), feed composition, and separation requirements.
- Separation processes use two primary mechanisms for performing the separation: partitioning between phases (equilibrium); or relative motion of various chemical species (rate).

- The selectivity of any separation is improved as the value of the separation factor is increased.
- Separation processes require a property difference and a separating agent.
- There are two types of separating agents: mass; and energy.
- The extent of separation depends on the size of the property difference and the specific separating agent(s) used.
- For separation processes, countercurrent operation is usually preferred over cocurrent operation because, in the former, the driving force is maintained at a larger value along a boundary.

2.13 Questions

2.1 Why is the concept of a unit operation useful?

2.2 Why is a complexing agent useful that can selectively and reversibly react with the solute of interest?

2.3 Describe why a difference in a property is needed for a separation.

2.4 What questions would you ask to evaluate various separation processes for a given situation?

2.5 Membranes separate species based upon differences in molecular size, a parameter which is often comparable to molecular weight. Why do you think it is difficult to partition air into pure oxygen and nitrogen streams using membranes?

2.6 Henry's Law constant is equal to the ratio of the vapor pressure of a species over the operating system pressure. If the number of stages required in a separation decreases with increasing Henry's Law constant, how can the Henry's Law constant be increased (i.e., of what thermodynamic variable is vapor pressure a function)?

2.7 Laboratory experiments were performed to assess the feasibility of separating ethylene from ethane. It was determined that the equilibrium solubilities of ethylene and ethane in an acidic copper(I) aqueous solution were similar. The rates of uptake of the two gases into the aqueous solution were measured independently and it was found that the rate of ethylene absorption is several times greater than that of ethane. Is this an example of an equilibrium- or a rate-controlled separation and why?

2.8 In membranes, the flux, or productivity, increases with increasing membrane pore size. Selectivity increases when the pore size is much smaller than the molecular diameter of the larger species in a binary mixture, but slightly larger than that of the smaller species. Explain the tradeoff between productivity and selectivity in terms of membrane pore sizes.

2.9 Describe circumstances in which steam can be either an energy-separating agent or a mass-separating agent.

2.10 List the two possible ways a separating agent obtains a different component distribution between two phases. Which of these applies to distillation, and why?

2.11 List the four ways in which separating agents generate selectivity. Which of these applies to (i) distillation, (ii) membrane separations, and (iii) adsorption?

2.12 Explain how an MSA used in reversible chemical complexation differs from that used in adsorption. What is the advantage of reversible chemical complexation over non-complexing separation processes?

2.14 Problems

2.1 Imagine that the height of an absorption column varies directly as the natural logarithm of the ratio of the exiting and entering contaminant concentration:

$$\text{height} = f[\ln(C_{\text{exit}}/C_{\text{enter}})].$$

To reduce a contaminant to 10% of its initial concentration, a column 3 meters tall is needed. For a similar separation, how tall must the column be to reduce the contaminant to 5% of its initial concentration?

2.2 The diameter of a distillation column operating at a specified gas velocity varies with the square root of the volumetric throughput (ft^3/s):

$$\text{diameter} = f(\text{throughput}^{1/2}).$$

If the throughput triples and the velocity remains the same, by what factor must the column diameter increase or decrease? For a column with a throughput of $100\,\text{ft}^3/\text{min}$, the diameter is 3.6 feet. A new column with a throughput of $260\,\text{ft}^3/\text{min}$ is to be built for a similar separation. What must the diameter be?

2.3 A Langmuir isotherm for an adsorption experiment shows that X, the amount of solute sorbed per weight sorbate, is 0.085 when the solute concentration in the fluid phase is 0.05. If the Langmuir constant K for the experimental conditions is 1.3, apply the concept of a unit operation to determine the value of X when the solute concentration is 0.1. Plot the Langmuir isotherm for C values of 2, 4, 6, 8, and 10. What does the plot tell you about the amount of solute absorbed vs the fluid phase concentration?

2.4 The list (a) to (r) on the facing page is a set of separation items and issues where a problem is identified and a solution implemented. For the chosen item:
(i)   State the separation objective.
(ii) Using the criteria for choosing a separation process, explain why this solution was chosen.
(iii) Describe the separation method(s) used to achieve this objective.
(iv) Provide data, calculations, flow diagrams, schematics, etc., to determine the ability of the separation method to achieve its objective.

(a)   Medical gloves

(b)   Medical breathing filters

(c)   Filters for household drinking-water faucets

(d)   Domestic drinking-water softeners

(e)   Water purification pumps for camping

(f)   Water purifiers for travel

(g)   Filter for drinking-water bottles

(h)   Recycle motor oil

(i)   Recycle freons or anti-freeze

(j)   VOC emissions reduction from dry cleaners

(k)   Decaffeinated coffee (solvent substitution)

(l)   Recovery and recycle of de-icing fluid at airports

(m)   Organics discharge reduction from brew pubs

(n)   Reduced water-phase discharge from car-wash facilities

(o)   Reduced gas- or water-phase emissions from a power plant

(p)   Metal membranes for water treatment

(q)   Waste treatment in space

(r)   Organics discharge reduction from gas stations

3

Separations analysis fundamentals

A journey of a thousand miles starts with a single step.

– ANCIENT PROVERB

As with most engineering evaluations, mass and energy balances must be used as part of the analysis of a particular separation process. Prior to an understanding of a specific separation process, an understanding of the basis for evaluation of these processes is required. Since separation processes involve separation of mass, the focus will be on the use of mass balances and mass transfer for analysis. This chapter, therefore, focuses both on mass balances, specifically as applied to separations, and on the fundamental concepts of mass transfer that are essential to comprehension of equilibrium- and rate-based processes.

3.1 Objectives

1. Write a mass balance in words.
2. Define a macroscopic and microscopic balance.
3. Define an overall and component mass balance.
4. State the basis for equilibrium- and rate-based process analysis.
5. Apply mass balances to analysis of equilibrium- and rate-based processes.
6. For equilateral and right triangle three-component equilibrium diagrams, identify:
   (a) the two-phase region;
   (b) tie-lines.
7. Be able to apply the lever-arm rule to find:
   (a) the mixing point (M) on a counter-current cascade;
   (b) the flowrates of streams (instead of using a mass balance).
8. Write the equations that describe the Langmuir and Freundlich isotherms and be able to determine the constants from batch-type adsorption test results.

9 Apply the graphical method (stepping off stages) to determine the number of equilibrium stages needed for a given separation requirement.
10 Calculate the number of equilibrium stages using the Kremser equation.
11 Calculate the minimum number of equilibrium stages using the Underwood equation and explain the physical significance.
12 Discuss the effect of stage efficiency on the number of required stages.
13 List sources of mass transfer resistance.
14 Calculate the overall mass transfer coefficient from the resistances in each phase.

3.2   Basic description of mass balances

A **mass balance** is nothing more than an accounting of material. Material balances can be written for a specific component or they can account for all mass in a system. A simple analogy to a mass balance would be an analysis of a personal financial budget. The overall balance would evaluate the total money received and spent during the period of evaluation, monthly, for example. An analysis of a specific item would be equivalent to a component balance. For example, a certain amount of money would be budgeted for certain items such as food, utilities, entertainment, etc. Each of these represents one component of the overall balance.

To perform a balance, a control volume (C.V.) must be identified first to isolate the system and the surroundings with respect to the balance. The surface of this control volume, as shown in Figure 3.1, is termed the control surface. The control volume is chosen to isolate the volume for evaluation, such as a complete separation process or just a specific portion. The control surface is chosen to identify the flow of mass into or out of the control volume. An **open** (or **flow**) **system** is one in which material is transferred across the control surface, that is, enters the C.V., leaves the C.V., or both. A **closed** (or **batch**) **system** is one in which there is no such transfer <u>during the time interval for evaluation</u>. Consider a vessel in which water is added that contains some dissolved organics and activated carbon. If the equilibration of the liquid and solid phases is of primary interest, then the system can be treated as a closed system if the vessel is selected as the C.V. If

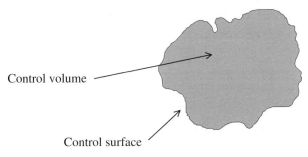

Control volume

Control surface

Figure 3.1 Generic control volume.

the amounts of water and/or activated carbon used over some time period (one month, for example) as well as the separation performance of a given batch are of primary interest, then the C.V. would be treated as an open system.

For the material presented in this book, the control volume is fixed in space (distillation columns usually don't move very much!). Mass can flow across the control surface. The conditions at the surface will be used in the balances as described below.

A component mass balance in words is:

The <u>net</u> rate at which the mass of a component enters the control volume (mass into minus mass out of the C.V.) plus the rate of mass of this component generated within the C.V. (i.e., chemical reactions) equals the rate of change of the mass of this component (i.e., accumulation) within the C.V. with respect to time.

$$\text{Accumulation} = \text{Input} - \text{Output} + \text{Generation}. \tag{3.1}$$

Accumulation is a positive quantity if there is an increase in the mass of the specific component with respect to time. A negative value implies a decrease. Likewise, the generation of a specific component by a chemical reaction would be a positive term while consumption in a chemical reaction would be a negative term.

A simple example at this point will illustrate the balance concept.

Example 3.1

Problem:

In a given year, 2,000 people moved to Boulder, CO, 500 people moved out, 1,500 were born and 1,000 died. Write a balance on the population of the city for this period. Is this a component or total balance?

Solution:

Accumulation = Input − Output + Generation

Input = 2,000

Output = 500

Generation = 1,500 − 1,000.

Substituting:

Accumulation = 2,000 − 500 + 1,500 − 1,000 = 2,000.

For that year, there was a net <u>increase</u> of 2,000 people. This is a total balance on the population of the city. If the number of women were selected for the balance, then this would be a component balance of the total population.

Mass balances for separation processes can refer to:

(a) Total mass (mass is normally used for liquid and solid flowrate measurements);

(b) total moles (gas flow);

(c) mass of a chemical compound (component balance);

(d) moles of a chemical compound (chemical reactions are described in terms of moles, not mass);

(e) mass of an atomic species (carbon balance, for example);

(f) moles of an atomic species (appropriate if there are chemical reactions).

A mass balance does not apply directly to volume. If the densities of the materials entering into each term are not the same, or mixing effects occur, then the volumes of the material will not balance. Also, chemical reactions can result in a change in the number of moles in the system (H_2 plus O_2 reacting to form H_2O is one example). For gas-phase reactions in which pressure and temperature remain constant, the volume can change (refer to the ideal gas law). Non-constant volumes can also apply to liquid-phase systems. For example, one classic mixing experiment is to mix equal volumes of alcohol and water. The resulting solution does not have twice the volume.

There are assumptions which eliminate certain terms in a mass balance.

1 Steady state. This assumption implies that there is no change with respect to time within the C.V. There may be changes with position. The result of this assumption is that the accumulation term is zero.

2 No chemical reactions occur within the control volume, therefore there is no generation.

Material balances can be either microscopic or macroscopic. Which to use is primarily dictated by the type of information desired. Figure 3.2 illustrates a hybrid process containing a distillation column and a membrane. The symbols indicate the different flowrates in the system. This process will be used to illustrate the two types of balances.

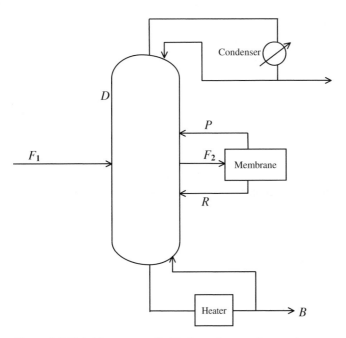

Figure 3.2 Hybrid process: a distillation column and a membrane.

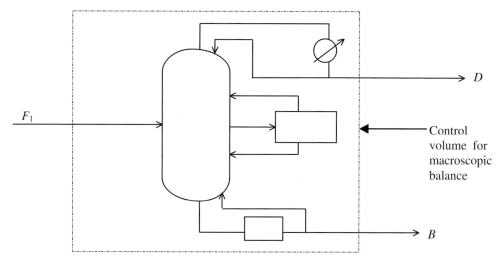

Figure 3.3 Control volume for macroscopic balance.

A macroscopic balance is used when global information is sufficient, i.e., no information at each point within the control volume is required. This type is used when the concentration profiles within the C.V. are not of interest, but the values associated with the masses and component concentrations that enter and leave the C.V. are of interest. This balance is sometimes described as an integral balance since any effects of position within the C.V. are averaged and only global values are evaluated. The system is treated as a "black box." In Figure 3.3 the composition and flowrate of streams F_1, D and B are of interest.

A microscopic (or differential) balance is used when information at each point within the control volume is needed. Each term of the balance equation is then a **rate** (rate of input, rate of generation, etc.). This type of balance is based on one or more differential equations that are solved to obtain the concentration profiles. The control volume in Figure 3.4 could be used if one were interested in the concentration profile across the membrane at each axial position of the membrane.

3.3   Degrees of freedom analysis

An important criterion in the solution of separation problems is determining that there is sufficient information to solve the problem. For any problem, there will be a certain number of independent variables (V) and a number of independent equations (E). The degrees of freedom (DF) or the number of independent variables that one needs to specify is

$$DF = V - E. \tag{3.2}$$

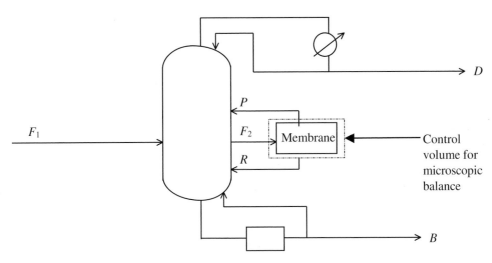

Figure 3.4 Control volume for microscopic balance.

If the actual number of variables specified (values given) is larger than DF, the system is said to be over-specified. Similarly, specifying less variables than DF causes the system to be under-specified. In both situations, a unique solution for the problems could not be obtained.

Some examples of variables would be composition, temperature, pressure, heat load, and flowrate of a given phase. If a phase composition is specified as mole fractions, then the weight fractions would not be a separate set of independent variables since they both measure composition and are directly related to each other. In other words, given the mole fractions and the molecular weights of each component (which are constants), the weight fractions can be calculated.

Some examples of equations are mass balances, energy balances, equilibrium relationships, and equalities (identities) of the temperature and/or pressure of two phases. If the system contains C components, then there are C independent mass balances. One example would be a mass balance for each component. In this case, the total mass balance is not independent since it is just the sum of the component balances.

Let's do some examples to illustrate this approach.

Figure 3.5 illustrates a batch two-phase system (vapor and liquid) in thermodynamic equilibrium. As before, y_i, x_i are the mole fractions of component i in the vapor and liquid phase, respectively. It is assumed that there are C components and each is present in both phases. The system has a constant temperature (T) and pressure (P).

The number of independent variables (V) is:

$2C$	component mole fractions (C components in each phase and their mole
T	fractions will normally be different in each phase)
P	

$\overline{2C + 2}$

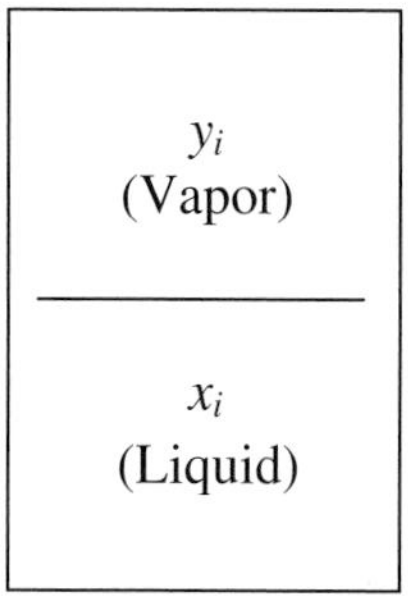

Figure 3.5 A batch two-phase system in thermodynamic equilibrium.

The number of independent equations (E) is:

2   (sum of mole fractions equals unity in each phase)

C (or similar thermodynamics equilibrium relationship for each component)

$\overline{}$

$C + 2$

$$DF = V - E = 2C + 2 - (C + 2) = C.$$

So, if the C component mole fractions in one phase are specified, one can uniquely solve for the other variables.

Now, let's expand the previous example to the case of $\mathcal{P}$ phases.

The number of independent variables (V) is

$\mathcal{P}C$ (component mole fractions)

T

P

$\overline{\phantom{\mathcal{P}C+2}}$

$\mathcal{P}C + 2$

The number of independent equations (E) is:

$\mathcal{P}$ (sum of mole fractions equals one in each phase)

$C(\mathcal{P} - 1)$ (equilibrium relationships. Note that in previous example we had two phases and $C\,(2 - 1)$ equilibrium relationships. Another way to think of this sum is that we start with the mole fraction in one phase and use these relationships to calculate the mole fraction in the other ($\mathcal{P} - 1$) phases)

$\overline{\phantom{\mathcal{P} + C(\mathcal{P} - 1)}}$

$\mathcal{P} + C(\mathcal{P} - 1)$

$$DF = V - E = \mathcal{P}C + 2 - [\mathcal{P} + C(\mathcal{P} - 1)] = C - \mathcal{P} + 2.$$

This is the Gibbs phase rule. This result demonstrates that the Gibbs phase rule is a specific case of this analysis and should only be applied for this specific situation.

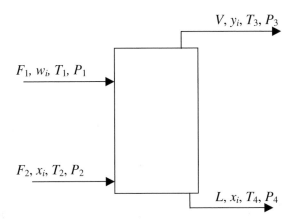

V, y_i, T_3, P_3

F_1, w_i, T_1, P_1

F_2, x_i, T_2, P_2

L, x_i, T_4, P_4

Figure 3.6 A non-adiabatic two-phase equilibrium-based process.

The above examples involve only intensive (not a function of process volume, such as density) variables. The analysis can be expanded to include extensive (total flowrate, total heat load, for example) variables.

Figure 3.6 shows a non-adiabatic (heat is either lost to or gained from the environment surrounding the control volume) two-phase equilibrium-based process. There are two feed streams and two exit streams. The exit streams are in thermodynamic equilibrium.

The number of independent variables is:

$4C$	(component mole fractions)
4	(stream flowrates)
$4T$	
$4P$	
Q	(net heat exchange (non-adiabatic process))

$4C + 13$

The number of independent equations is:

4	(sum of mole fractions equals unity)
C	(component mass balances)
1	(energy balance)
C	(phase equilibrium relationships)
1	(temperature equality (exit T are equal))
1	(pressure equality (exit P are equal))

$2C + 7$

$$DF = V - E = 4C + 13 - (2C + 7) = 2C + 6.$$

To solve this problem uniquely, one would have to specify $2C + 6$ variables. One example would be to specify the composition, flowrate, T and P of the two feed steams. If the exit streams are *not* in thermodynamic equilibrium, then the phase equilibrium relationships cannot be used and the exit streams may not be at the same T and P.

Important additional point. The value of the variables specified has to be reasonable for the system. For example, if the feed stream for a two-phase (vapor and liquid) equilibrium-limited process is water-based, and the equilibrium pressure is specified as 1 atm and the temperature as 200 °C, there will not be a two-phase system at equilibrium.

3.4   Phase equilibrium

Phase equilibrium information characterizes partitioning between phases for a system and is important for describing separation processes. For equilibrium-limited processes, these values dictate the limits for separation in a single stage. For mass transfer-limited processes, the partitioning between phases is an important parameter in the analysis. The data can be presented in tabular form. But this approach is restricted in application, since an analysis typically requires phase equilibrium values that are not explicitly listed in the table. So, graphical representation and computational methods are usually more useful.

3.4.1   Vapor–liquid equilibrium (VLE)

For binary systems, graphical representation is typically plotted (Figure 3.7(a)) as the vapor-phase mole fraction of the more volatile component (y_A) vs liquid-phase mole fraction (x_A). The $y = x$ diagonal line is included for reference. The data are usually plotted for a constant total pressure.

Figure 3.7(b) illustrates the presence of an azeotrope (a point at which $y_A = x_A$). The presence of an azeotrope is important since it represents a limiting value in distillation.

An alternative graphical method, shown in Figure 3.8, is a T–x–y phase diagram. Again, the total pressure is constant. To use the graph, select a temperature, draw a horizontal

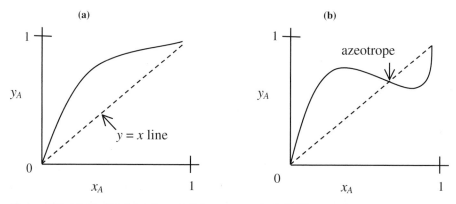

Figure 3.7 (a) Equilibrium line (solid) on a typical VLE x–y diagram; (b) graphical representation of an azeotrope in a VLE system.

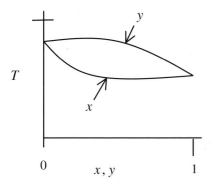

Figure 3.8 *T–x–y* graphical representation of a VLE system.

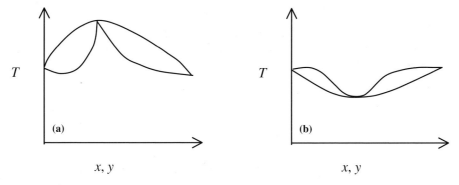

Figure 3.9 (a) Maximum boiling-point azeotrope; (b) minimum boiling-point azeotrope.

line at that value, and read the corresponding values of x_A and y_A from the respective curves.

Two types of azeotrope, maximum boiling point (Figure 3.9(a)) and minimum boiling point (Figure 3.9(b)), can be represented on this type of diagram.

If the VLE data can be represented by a linear relationship, $y = Kx$, a graph of K vs total pressure can be one method to represent the data.

3.4.2 Calculation methods

K-values

A K-value K_i, can be defined for each component (i) in the vapor or liquid phase:

$$K_i \equiv \frac{y_i}{x_i}.$$ (3.3)

This allows for easy use in calculations of separation processes where phase equilibrium data are needed. Historically, this approach was used prior to the advent of personal computers when calculations involving linear terms were much simpler to implement.

Use of activity coefficients (γ_i)

Most systems cannot be described by $y_i = K_i x_i$ for a wide range of x_i values. This equation only characterizes linear systems and is called an ideal relationship. For non-ideal systems this relationship is modified by an activity coefficient, γ_i, for each component.

At low pressures (up to a few atmospheres):

$$\frac{y_i}{x_i} = \gamma_i \frac{p_i^0}{P}, \tag{3.4}$$

[note: fugacity coefficients are assumed to cancel each other (or nearly) and Poynting corrections are assumed to be unity (or close)]

where y_i = vapor mole fraction of component i

$\qquad x_i$ = liquid mole fraction of component i

$\qquad \gamma_i$ = activity coefficient of component i

$\qquad p_i^0$ = vapor pressure of component i

$\qquad P$ = total pressure.

Vapor pressure is typically predicted by an Antoine equation:

$$\log_{10}\left[p_i^0\right] = A - \frac{B}{T + C}, \tag{3.5}$$

where A, B, C = constants

$\qquad T$ = temperature, °C.

Activity coefficients are generally predicted by one of the Wilson, UNIQUAC, NRTL, or van Laar methods. The Wilson and UNIQUAC methods are presented briefly here. Most chemical engineering thermodynamics textbooks have a section on phase equilibria that can provide more detailed descriptions. The Wilson equation [1] is only used with miscible fluids. For highly non-ideal fluids and for systems in which liquid–liquid splitting occurs, the NRTL method is applicable [2]. When no experimental data are available, the UNIQUAC method can be used [3, 4].

Wilson equation

$$\ln \gamma_i = -\ln\left(\sum_{j=1}^{m} x_j \Lambda_{ij}\right) + 1 - \sum_{k=1}^{m} \frac{x_k \Lambda_{ki}}{\sum_{j=1}^{m} x_j \Lambda_{kj}}, \tag{3.6}$$

where

$$\Lambda_{ij} = \frac{V_j^L}{V_i^L} \exp\left[-\frac{\lambda_{ij} - \lambda_{ii}}{RT}\right]; \tag{3.7}$$

and

$$\Lambda_{ii} = \Lambda_{jj} = 1. \tag{3.8}$$

Here, V_i^L = molar volume of pure liquid component i

λ_{ij} = interaction energy between components i and j, $\lambda_{ij} = \lambda_{ji}$

T = absolute temperature, in Kelvin, K

R = gas law constant.

UNIQUAC equation

$$\ln \gamma_i = \ln \gamma_i^C + \ln \gamma_i^R, \tag{3.9}$$

where

$$\ln \gamma_i^C = \ln \frac{\varphi_i}{x_i} + \frac{z}{2} q_i \ln \frac{\vartheta_i}{\varphi_i} \sum_j x_i l_j \tag{3.10}$$

and

$$\ln \gamma_i^R = q_i \left[1 - \ln \left(\sum_{j=1}^m \vartheta_j \tau_{ji} \right) - \sum_{j=1}^m \frac{\vartheta_j \tau_{ij}}{\sum\limits_{k=l}^m \vartheta_k \tau_{kj}} \right], \tag{3.11}$$

with

$$l_i = \frac{z}{2} (r_i - q_i) - (r_i - l), \qquad z = 10 \tag{3.12}$$

$$\tau_{ji} = \exp \left[-\frac{u_{ji} - u_{ii}}{RT} \right], \qquad \tau_{ii} = \tau_{jj} = 1. \tag{3.13}$$

Here, q_i = area parameter of component i

r_i = volume parameter of component i

u_{ij} = parameter of interaction between components i and j, $u_{ij} = u_{ji}$

z = coordination number

γ_i^C = combinatorial part of activity coefficient of component i

γ_i^R = residual part of activity coefficient of component i

$\vartheta_i = q_i x_i / \sum_j q_j x_j$ = area fraction of component i

$\varphi_i = r_i x_i / \sum_j r_j x_j$ = volume fraction of component i.

3.4.3 Triangular diagrams

Extraction, to be covered in Chapter 5, is a three-component process in which a solute is transferred from a diluent into a solvent. Binary phase diagrams at constant temperature and pressure used in analyzing distillation are insufficient to characterize a three-component system. Triangular diagrams (at constant T and P) represent the equilibrium phase behavior of a three-component partially miscible system. The equilateral triangular diagram is easy to read, but somewhat difficult to draw. The right triangle diagram is much easier to produce, but equilateral diagrams are often found in literature so it is important to understand how to read them.

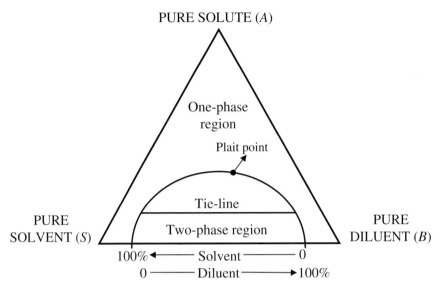

PURE SOLUTE (*A*)

One-phase
region

Plait point

Tie-line

**PURE
SOLVENT (*S*)**

Two-phase region

**PURE
DILUENT (*B*)**

100% ◄──── Solvent ──── 0
0 ──── Diluent ────► 100%

Figure 3.10 Three-component phase equilibrium diagram.

The three apexes of the equilateral triangle (Figure 3.10) represent each of the three pure components. In this figure, pure solvent (*S*) is the lower left corner, pure solute (*A*) is the top apex, and pure diluent (*B*) is the lower right corner. The curved saturation line in the interior of the triangle forms a solubility envelope, under which lies the two-phase region of immiscible solvent and diluent. At any point above this line the system is completely miscible and forms only one phase. Extraction is not an effective means of separation in this one-phase region, because the solvent and diluent must be immiscible in order for partitioning of the solute from diluent to solvent and subsequent separation to occur. If a system is operating in this one-phase regime, a simple addition of solvent can dilute the mixture into the two-phase region. The plait point is similar to an azeotrope for vapor–liquid equilibrium because it is the point at which the two phases have identical compositions. The saturation line to the left of the plait point is the extract (solvent-rich) phase, and the equilibrium compositions of the extract phase are given by points on this line. The saturation line to the right of the plait point is the raffinate (diluent-rich) phase, and the equilibrium compositions of the raffinate phase are given along this line. It should be noted that one corner could represent any of the pure components, and some triangle diagrams will be plotted differently than others.

Any point inside the two-phase region represents a mixture which will separate into two phases along a tie-line. The compositions of these two phases lie on the saturation line at the ends of the tie-line. The most important thing to remember about tie-lines is that they show the compositions of two phases *which are in equilibrium with each other*. The data used to plot the saturation line and the tie-lines are usually obtained experimentally; there are no simple equilibrium relationships or equations as in Henry's Law used in distillation analyses. A tie-line that slopes down from the extract line to the raffinate line

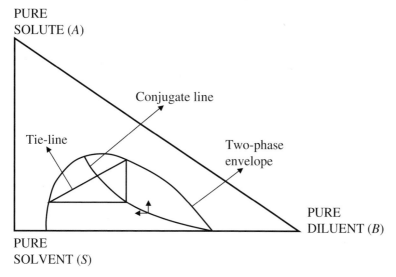

PURE
SOLUTE (*A*)

Conjugate line

Tie-line

Two-phase
envelope

PURE
DILUENT (*B*)

PURE
SOLVENT (*S*)

Figure 3.11 Construction of a tie-line from the conjugate line.

indicates a favorable separation. In these systems, the equilibrium solute concentration in the extract is greater than that in equilibrium with the raffinate. Hence, less solvent is needed to remove the solute from the diluent. Separation is still possible in systems in which tie-lines slope up from the extract line to the raffinate line. However, more solvent than in the prior case will be required to achieve the separation.

The right triangular diagram (Figure 3.11) is similar to the equilateral triangular diagram; each corner represents one of the pure components.

There are two additional ways that equilibrium (tie-line) data can be shown on a triangular diagram. The first, presented in Figure 3.11, is a conjugate line: a vertical line from any point on the conjugate line gives one end of the tie-line (on the saturation line), and a horizontal line from the same point will locate the other end of the tie-line. An equilibrium curve may also be used to locate tie-lines on a triangular diagram.

Conjugate lines and equilibrium curves can avoid cluttering up the diagram by allowing only the tie-lines of interest to be drawn. They can be used on equilateral triangle diagrams as well as on right triangle diagrams, but care should be taken in transcribing equilibrium data from an *x*–*y* diagram to an equilateral triangle diagram which will not have the same axes.

Figure 3.12 illustrates an alternate method by which to generate a triangular diagram.

Regardless of the type of triangle diagram that is being used, remember that the weight fractions of each phase must add up to one:

$$x_A + x_B + x_S = y_A + y_B + y_S = 1.$$

This is a useful check to confirm that a triangle diagram is read correctly.

It is also important to know that as temperature increases, miscibility almost always increases and the two-phase region shrinks, as shown in Figure 3.13. This dependence

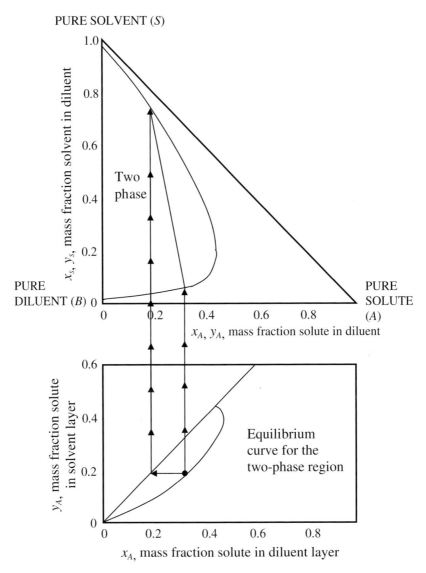

Figure 3.12 Construction of a tie-line from a single point on the equilibrium curve.

can be used to alter the phase equilibrium for any system to obtain a more favorable separation.

3.4.4 Adsorption isotherms

In adsorption a solute (sorbate) is transferred from a fluid (gas, vapor or liquid) to a porous solid adsorbent phase. The driving force for all adsorptive processes lies in the initial departure from thermodynamic equilibrium between the fluid and solid phases. The rate

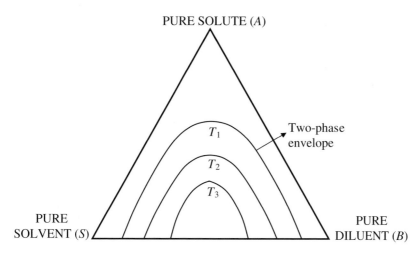

Figure 3.13 Effect of temperature on solubility envelope ($T_3 > T_2 > T_1$).

of mass transfer is limited by either the rate of diffusion through the fluid phase or the rate of sorption onto or into the sorbent. The sorption/desorption steps are normally considered to be rapid relative to the diffusion steps. So, equations describing the equilibrium between the fluid and solid phase (at constant temperature) have been developed; these equations are called isotherms. Isotherms allow one to predict the capacity of a given sorbent for a particular sorbate as a function of sorbate concentration in the fluid phase. More complete descriptions and derivations are available elsewhere [5–7].

Single-component isotherms

Isotherms are derived from equilibrium measurements of mass of sorbate adsorbed for a given mass of sorbent as a function of sorbate concentration in the fluid phase in contact with the sorbent at constant temperature. Based on the shape of the resulting curve, the measurements can be fitted to an equation. Figure 3.14 illustrates the various types of isotherms and their classification by Roman numeral. Types I and II are the most common. A Type I isotherm is representative of sorbents with very small pore size which lead to unimolecular adsorption. Types II and III are representative of sorbents with a pore-size distribution such that multilayer sorption (and even capillary condensation) are possible. Types IV and V are characterized by a hysterisis loop. The amount adsorbed at equilibrium is different at a given solute concentration in the fluid phase depending on whether the fluid phase concentration is increasing or decreasing. This effect implies that it is difficult to desorb the solute once it is adsorbed within the concentration range of the hysterisis loop (can you explain why?). Note that Types IV and V isotherms have similar shape to Types II and III respectively during sorption.

Once the measurements have been obtained, the data can be fitted to an isotherm as stated above. The model equations can be derived based on various assumptions concerning the accessibility of adsorption sites, the number of molecules that can be adsorbed per site

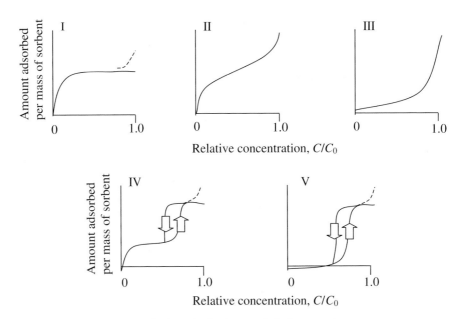

Figure 3.14 The five types of adsorption isotherms. C_0 is the saturation concentration of solute in fluid phase.

and the adsorption energy distribution for the sites. These derivations can be found in various references [4–6]. The resulting quality of fit can then provide information about the adsorption process for the particular sorbate–sorbent pair. The most commonly used isotherm equations will be presented below.

Langmuir isotherm

Langmuir's theory for deriving an isotherm is a kinetic one, assuming the adsorption system is in dynamic equilibrium, where the rate of adsorption is equal to that of desorption. The Langmuir isotherm, described by the following equation, is still the most useful for data correlation.

$$\frac{x}{m} = q = \frac{aKC_e}{1 + KC_e} \tag{3.14}$$

where x = mass of solute adsorbed

m = mass of adsorbent

q = mass ratio of the solid phase – mass of adsorbed solute per mass of adsorbent

C_e = equilibrium concentration of solute (mass/volume)

a = mass of adsorbed solute required to completely saturate a unit mass of adsorbent (constant)

K = experimental constant.

Some values of these constants are provided in Table 3.1.

Table 3.1 *Selected values of the Langmuir-isotherm parameters (Equation 3.14) for adsorption of various organic compounds on PAC [8].*

Compound	a (mmol/g)	K (L/mmol)	Compound	a (mg/mg)	K (L/mg)
i-propanol	0.79	0.123	Aniline	0.108	0.177
n-propanol	0.92	0.158	Benzyl alcohol	0.158	0.176
i-butanol	1.04	0.482	Benzaldehyde	0.210	0.170
n-butanol	1.06	0.954	2-naphthol	0.345	0.643
n-pentanol	2.85	0.858	*o*-toluidine	0.186	0.256
n-hexanol	3.21	1.35	*p*-toluidine	0.186	0.256
Propionaldehyde	1.21	0.107	*o*-anisidine	0.240	0.417
Butyraldehyde	1.19	0.745	*m*-anisidine	0.185	0.338
Ethyl acetate	1.54	0.531	*p*-anisidine	0.248	0.505
i-propyl acetate	1.66	0.954	Anisaldehyde	0.305	0.468
Propyl acetate	7.13	0.153	Salicylaldehyde	0.240	0.417
Butyl acetate	10.7	0.166	Vanillin	0.375	1.110
Methyl ethyl ketone	1.00	0.526	Pyridine	0.055	0.154
Diethyl ketone	1.67	0.911	2-methylpyridine	0.106	0.262
Methyl i-butyl ketone	1.95	1.44	*o*-cresol	0.240	0.417
			2-chlorophenol	0.272	0.405
			Nitrobenzene	0.310	0.230
			o-methoxyphenol	0.296	0.400
			Quinaldine	0.296	0.400
			Pyrrole	0.030	0.152
			Indole	0.240	0.417

The Langmuir isotherm is based on the following assumptions:

1 The adsorbent surface has a fixed number of identical individual "spaces" in which an adsorbate molecule can reside.

2 The adsorbent surface will accumulate only one layer of adsorbate molecules.

3 Reversible chemical equilibrium is assumed to exist.

When $KC_e \ll 1$, the Langmuir isotherm reduces to a linear form analogous to Henry's Law. Sorbate–sorbent pairs that display a Type I isotherm can use the Langmuir equation.

Example 3.2: determining Langmuir-isotherm constants

Problem:

Laboratory tests were conducted on an aqueous stream containing 50 mg/L phenol. Various amounts of powdered activated carbon (PAC) were added to four containers each containing 1 liter of the wastewater. When equilibrium was reached, the phenol concentration in each container was measured and is tabulated below. Determine the Langmuir-isotherm constants for this system and calculate the amount (mass of carbon required) required to reduce the phenol concentration to 0.10 mg/L for a 1-liter sample.

Results:

Container #	Carbon added (g)	Equilibrium phenol concentration (mg/L)
1	0.50	6.00
2	0.64	1.00
3	1.00	0.25
4	2.00	0.08

Solution:

The Langmuir isotherm can be rewritten:

$$\frac{C_e}{(x/m)} = \frac{1}{aK} + \frac{1}{a}C_e, \tag{3.15}$$

so that a plot (Figure 3.15) of $C_e/(x/m)$ vs C_e is a straight line (remember that x is the amount of phenol adsorbed, which is the equilibrium amount subtracted from the original amount). A linear regression of the data gives a straight line with a slope of 11 and an intercept of 2.2. Solving for a and K gives $a = 9.1 \times 10^{-2}$ g phenol/g PAC and $K = 4.9$ L/mg. [Remember the physical meaning of the constant a: it will take 91 mg of adsorbed phenol to completely saturate 1 g of carbon.]

Once the constants are determined, the amount of carbon required to obtain an equilibrium concentration of 0.10 mg/L phenol can be easily determined (be sure not to mix mg and g units):

mass carbon required = 170 mg for 1 liter solution.

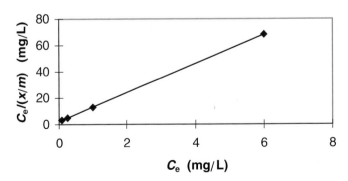

Figure 3.15 A plot of $C_e/(x/m)$ vs C_e, Example 3.2.

Freundlich isotherm

The Langmuir–Freundlich equation is

$$q = \frac{(KC_e)^{1/n}}{1 + (KC_e)^{1/n}}, \tag{3.16}$$

where q = mass ratio of the solid phase – mass of adsorbed solute per mass of adsorbent

C_e = equilibrium concentration of solute (mass/volume)

K, n = experimental constants (some values are provided in Table 3.2).

The Langmuir–Freundlich equation is similar in form to the Langmuir isotherm. When the adsorbate occupies n sites, it is modified to:

$$q = \frac{(BP)^{1/n}}{1 + (BP)^{1/n}},$$
(3.17)

where B is a constant and P is the total pressure.

This equation is the Langmuir–Freundlich isotherm; it differs from the Langmuir isotherm in two ways. First, it does not assume that the energy of adsorption is the same for all surface sites. In reality, the energy of adsorption will vary because real surfaces are heterogeneous. The isotherm expression tries to account for this by assuming that the frequency of sites associated with a free energy of adsorption decreases exponentially with increasing free energy. Second, it is assumed that each adsorbate occupies n sites.

When the term in the denominator is close to 1, $(BP)^{1/n} \ll 1$, the equation simplifies to the Freundlich isotherm:

$$\frac{x}{m} = q = KC_e^{1/n},$$
(3.18)

where x = mass of solute adsorbed

m = mass of adsorbent.

The Langmuir and Freundlich isotherms are the most common isotherms; it is likely that the constants in these isotherms for many sorbate–sorbent pairs have already been measured (see Tables 3.1 and 3.2, respectively).

Example 3.3: determining Freundlich-isotherm constants

Problem:

Treated and filtered wastewater can be recycled for use in irrigation. An important issue is odor removal prior to reuse. For a wastewater that has an initial concentration of 10 ppm of these compounds, the following test results are obtained when activated carbon is used to adsorb them:

Carbon added (mg/L)	0.0	0.4	1.0	6.0
Concentration (ppm)	10	6.9	4.5	1.5

Using the Freundlich isotherm, determine the minimum amount of activated carbon required to reduce the concentration to 0.20 ppm.

Solution:

A plot (Figure 3.16) of log(concentration adsorbed/mass carbon added) vs log(concentration at equilibrium) gives a straight line. A linear regression of the data

Table 3.2 *Selected values of the Freundlich-isotherm parameters (Equation 3.16) for adsorption of various organic compounds onto activated carbon [8].*

	K (mg/g)	1/n		K (mg/g)	1/n
Acenaphthelene	190	0.36	n-dimethylnitrosamine	6.8×10^{-5}	6.6
Acenaphthylene	115	0.37	2,4-dimethylphenol	78	0.44
Acrolein	1.2	0.65	Dimethyl phthalate	97	0.41
Acrylonitrile	1.4	0.51	4,6-dinitro-*o*-cresol	237	0.32
Aldrin	651	0.92	2,4-dinitrophenol	160	0.37
Anthracene	376	0.70	2,4-dinitrotoluene	146	0.31
Benzene	1.0	1.6	2,6-dinitrotoluene	145	0.32
Benzidine-dihydrochloride	110	0.35	1,2-diphenylhydrazine	16,000	2.0
3,4-benzofluoroanthene	57.0	0.37	Alpha-endosulfan	194	0.50
Benzo[k]fluoroanthene	181	0.57	Beta-endosulfan	615	0.83
Benzo[ghi]perylene	10.7	0.37	Endosulfan sulfate	686	0.81
Benzo[a]pyrene	33.6	0.44	Endrin	666	0.80
Alpha-BHC	303	0.43	Ethylbenzene	53	0.79
Beta-BHC	220	0.49	bis-(2-ethylhexyl)phthalate	11,300	1.5
Gamma-BHC (lindane)	256	0.49	Fluoranthene	664	0.61
Bromoform	19.6	0.52	Fluorene	330	0.28
4-bromophenyl phenyl ether	144	0.68	Heptachlor	1220	0.95
Butylbenzyl phthalate	1520	1.26	Heptachlor epoxide	1038	0.70
n-butyl phthalate	220	0.45	Hexachlorobenzene	450	0.60
Carbon tetrachloride	11.1	0.83	Hexachlorobutadiene	258	0.45
Chlorobenzene	91	0.99	Hexachlorocyclopentadiene	370	0.17
Chordane	245	0.38	Hexachloroethane	96.5	0.38
Chloroethane	0.59	0.95	Isophorone	32	0.39
bis-(2-chloroethoxy)methane	11	0.65	Methylene chloride	1.30	0.16
bis-(2-chloroethyl)ether	0.086	1.84	4,4'-methylene-bis-	190	0.64
2-cholorethyl vinyl ether	3.9	0.80	(2-chloroaniline)		
Chloroform	2.6	0.73	Naphthalene	132	0.42
bis-(2-chloroisopropyl)ether	24	0.57	Beta-naphthylamine	150	0.30
Parachorometa cresol	122	0.29	Nitrobenzene	68	0.43
2-chloronaphthalene	280	0.46	2-nitrophenol	101	0.26
2-chlorophenol	51.0	0.41	4-nitrophenol	80.2	0.17
4-chlorophenyl phenyl ether	111	0.26	n-nitrosodiphenylamine	220	0.37
DDE	232	0.37	n-nitrosodi-n-propylamine	24.4	0.26
DDT	322	0.50	PCB-1221	242	0.70
Dibenzo[a,h]anthracene	69.3	0.75	PCB-1232	630	0.73
1,2-dichlorobenzene	129	0.43	Pentachlorophenol	260	0.39
1,3-dichlorobenzene	118	0.45	Phananthrene	215	0.44
1,4-dichlorobenzene	121	0.47	Phenol	21	0.54
3,3-dichlorobenzidine	300	0.20	1,1,2,2-tetrachloroethane	10.6	0.37
Dichlorobromomethane	7.9	0.61	Tetrachloroethane	50.8	0.56
1,1-dichloroethane	1.79	0.53	Toluene	26.1	0.44
1,2-dichloroethane	3.57	0.83	1,2,4-trichlorobenzene	157	0.31
1,2-*trans*-dichloroethane	3.05	0.51	1,1,1-trichloroethane	2.48	0.34
1,1-dichloroethylene	4.91	0.54	1,1,2-trichloroethane	5.81	0.60
2,4-dichlorophenol	147	0.35	Trichloroethene	28.0	0.62
1,2-dichloropropane	5.86	0.60	trichlorofluoromethane	5.6	0.24
Dieldrin	606	0.51	2,4,6-trichlorophenol	219	0.29
Diethyl phthalate	110	0.27			

gives the slope and the intercept of a straight-line fit, and the constants n and K can be found. Taking logarithms of both sides of the isotherm gives:

$$\log \left(\frac{x}{m}\right) = \log K + \frac{1}{n} \log C_e.$$

The value of the slope in the plot is $1/n$, and the intercept is $\log(K)$. Therefore $n = 0.73$ (unitless) and $K = 0.91$ (ppm/mg · L carbon). Now that the constants are known, the isotherm can be solved for the mass/volume of carbon required to reduce the concentration to 0.2 ppm. Remember that x is not concentration, but concentration adsorbed (C_e is concentration at equilibrium). Therefore:

98 mg/L of carbon are required for a concentration of 0.2 ppm.

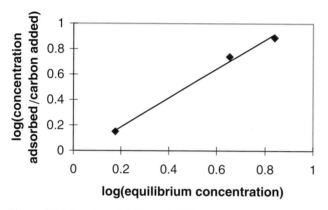

Figure 3.16 Log–log plot, Example 3.3.

Mixture isotherms

Langmuir has also considered the dissociative adsorption for the case of each molecule occupying two sites. In this case two sites are needed for both adsorption and desorption, and hence the rates are proportional to $(1 - X)^2$ and X^2 (where $X = q/a$) for adsorption and desorption, respectively. The resulting isotherm for gas sorption is:

$$q = \frac{(BP)^{1/2}}{1 + (BP)^{1/2}}, \tag{3.19}$$

where, as before, B is a constant and P is pressure. Knaebel [9] lists a number of single-component and mixture isotherms. Single-component and mixture equilibrium data are available (refer to references in Chapter 7, Adsorption).

3.5 Equilibrium-limited analysis

As stated previously, the degree of separation in an equilibrium-limited process is restricted to a single contact. So, this process is normally carried out in sequential stages to improve

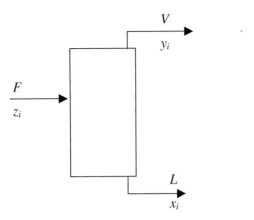

Figure 3.17 Continuous equilibrium flow system.

the separation. There are various staging modes and they will be described in more detail in later chapters on specific separation technologies. The main purpose here is to show a basic example for a single stage that will form the basis for analyses of various sequences of equilibrium stages.

3.5.1 One equilibrium stage

Figure 3.17 represents a continuous flow system where streams V and L are in thermodynamic equilibrium. This system can be evaluated by using thermodynamic equilibrium information with the appropriate number of mass balances. This will be illustrated first graphically and then analytically for a binary system. To analyze the system graphically, an equation must be obtained $y = f(x)$ from mass balances (operating line) and plotted on an x–y equilibrium diagram. The intersection of the mass balance and equilibrium line is the solution.

Component mass balance:

$$z_A F = x_A L + y_A V. \tag{3.20}$$

Total mass balance:

$$F = L + V \rightarrow L = F - V. \tag{3.21}$$

Combining and rearranging:

$$y_A = x_A \frac{[(V/F) - 1]}{V/F} + z_A \frac{1}{V/F}. \tag{3.22}$$

When $y_A = x_A$:

$$x_A = x_A \frac{[(V/F) - 1]}{V/F} + z_A \frac{1}{V/F} \rightarrow x_A = z_A. \tag{3.23}$$

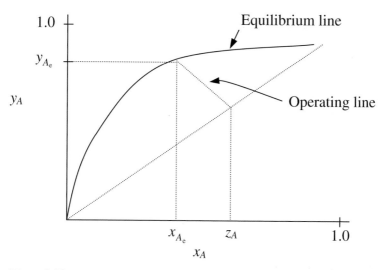

Figure 3.18 $x-y$ analysis of a single equilibrium-limited stage.

Therefore, the feed mole fraction z_A is known, since the operating line intersects the $y = x$ line at z_A. Then, the V/F ratio can be specified and the equilibrium values obtained, or vice versa. This is shown in Figure 3.18.

The same calculation can be performed analytically using an equation that relates y_{A_e} and x_{A_e}. For a constant relative volatility, α_{AB},

$$\alpha_{AB} = \frac{y_A/x_A}{y_B/x_B} = \frac{y_A/x_A}{(1 - y_A)/(1 - x_A)} \tag{3.24}$$

$$y_{A_e} = \frac{\alpha_{AB} x_{A_e}}{1 + x_A(\alpha_{AB} - 1)}. \tag{3.25}$$

With two equations and two unknowns one can solve for x_{A_e} and y_{A_e}.

An alternative graphical solution uses a $T-x-y$ phase equilibrium diagram. First, the mass balance equation is rearranged:

$$z_A = x_A \left(1 - \frac{V}{F}\right) + y_A \frac{V}{F} \tag{3.26}$$

$$\frac{z_A - x_A}{y_A - x_A} = \frac{V}{F}. \tag{3.27}$$

If the mole fractions are plotted on a $T-x-y$ diagram (Figure 3.19), V/F can be calculated from the ratio of line segments. This is an example of the lever-arm rule (described in to next section). Note that the mass balance could be rearranged to solve for L/V or L/F using the same approach of the ratio of the length of line segments. Note also that once T is fixed, the x_{A_e} and y_{A_e} are fixed. The ratio of flowrates V/F can be varied by changing z_A.

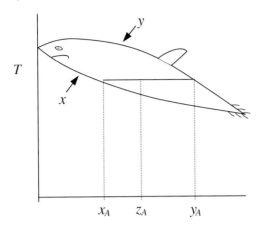

Figure 3.19 T–x–y phase equilibrium diagram for a single-stage process.

For liquid–liquid extraction systems, the approach is the same to analyze a single equilibrium-limited contact. A mass balance (operating) line is plotted on a phase equilibrium diagram. In this case, the phase equilibrium diagram is a ternary diagram.

3.5.2 Lever-arm rule

Each equilibrium stage in extraction has two distinct steps, mixing of the two phases for solute partitioning and separation of the phases to form exit phases with altered solute concentrations. The goal of analysis for any system is, as before, to determine the number of equilibrium stages required for a specified separation. The lever-arm rule described in this section provides a way of analyzing a tertiary system by breaking it down into mixing and separating steps.

The analysis begins with mixing of the solvent and diluent streams, as follows. Imagine two liquid streams (O and V) which may contain any or all of components A, B and C. These two streams are mixed to form a third stream, (F). The streams O, V and F may be single phase or two phase. Note that the control volume isn't necessarily an equilibrium-limited stage. The compositions and flowrates of the two feed streams are known (remember that for O: $x_A + x_B + x_C = 1$, for V: $y_A + y_B + y_C = 1$, and for F: $z_A + z_B + z_C = 1$). [*Do not be confused about the notation: the x's and y's are used to differentiate between the compositions of the two feeds, but the y's do not mean that stream V is vapor. Both feed streams are liquid.*]

Referring to Figure 3.20:

V = total mass (or flowrate) of V phase
O = total mass (or flowrate) of O phase
F = total mass (or flowrate) of F phase   Assumptions:
x_{Ai} = fraction of component A in O phase   1 no chemical reactions
y_{Ai} = fraction of component A in V phase   2 isothermal system
z_{Ao} = fraction of component A in F phase   3 system is at steady state.

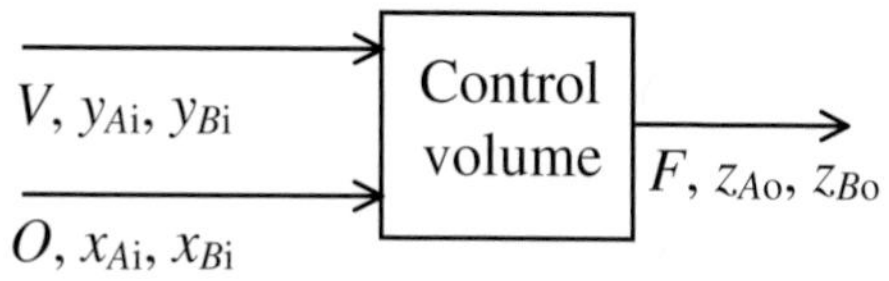

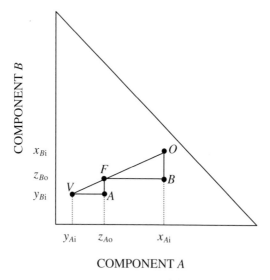

Figure 3.20 Control volume used to explain lever-arm rule.

Figure 3.21 Graphical representation of mass balances for a two-liquid feed stream on a right triangle diagram.

The mass balances for the system are represented graphically on a right triangular diagram (Figure 3.21) [2].

Analytically, there are three independent mass balances for this system:

Overall: $\qquad V + O = F$

Component A: $V y_{Ai} + O x_{Ai} = F z_{Ao}$

Component B: $V y_{Bi} + O x_{Bi} = F z_{Bo}.$

Combining these equations, it can be shown that the concentrations of the mixed stream F are then:

$$z_{Ao} = \frac{V y_{Ai} + O x_{Ai}}{V + O} \qquad (3.28)$$

$$z_{Bo} = \frac{V y_{Bi} + O x_{Bi}}{V + O}. \qquad (3.29)$$

The mass balances can be combined into two equations to eliminate F. The result proves that the points V, O and F are colinear as pictured in Figure 3.21,

$$\frac{V}{O} = \frac{z_{Ao} - x_{Ai}}{y_{Ai} - z_{Ao}} \quad \text{and} \quad \frac{V}{O} = \frac{z_{Bo} - x_{Bi}}{y_{Bi} - z_{Bo}}. \qquad (3.30)$$

Thus, if the inlet and outlet compositions are known, then V/O can be determined. For an extraction process, this corresponds to the required solvent/diluent flow ratio. The ratios O/F and V/F could be determined with the appropriate substitution in one of the component balances. [Remember: these ratios are determined by mass balances around the control volume. Nothing has been stated about thermodynamic equilibrium within the control volume.]

Setting these equations equal to each other and rearranging gives:

$$\frac{z_{Bo} - x_{Bi}}{z_{Ao} - x_{Ai}} = \frac{y_{Bi} - z_{Bo}}{y_{Ai} - z_{Ao}}. \tag{3.31}$$

The left-hand side of this equation is the slope from F to O, and the right-hand side is the slope from F to V. The equation itself is in the three-point form of a straight line. Hence, the points (z_{Ao}, z_{Bo}), (x_{Ai}, x_{Bi}), and (y_{Ai}, y_{Bi}) are all on a straight line.

Using the similar triangles in the figure:

$$\frac{\overline{FO}}{\overline{VF}} = \frac{\overline{FB}}{\overline{VA}} = \frac{z_{Bo} - x_{Bi}}{y_{Bi} - z_{Bo}}. \tag{3.32}$$

The right-hand side of this equation was already shown to be equal to V/O, therefore:

$$\frac{V}{O} = \frac{\overline{FO}}{\overline{VF}} = \text{lever-arm rule}. \tag{3.33}$$

When applied to extraction problems, the two feed streams V and O are equivalent to the incoming feed and solvent streams. The stream F would represent a two-phase mixture, which would separate into the raffinate and extract phases. The component A is usually the solute and the component B is usually the diluent, although the lever-arm rule will work no matter how the axes of the diagram are arranged. *When solving extraction problems graphically, it is really useful to remember equations:*

$$z_{Ao} = \frac{V y_{Ai} + O x_{Ai}}{V + O} \qquad \text{or} \qquad z_{Bo} = \frac{V y_{Bi} + O x_{Bi}}{V + O}$$

that help to plot the mixing point without having to re-solve the mass balances. Either one of these co-ordinates is sufficient to plot the mixing point, since it is already known to be colinear with the points of the two inlet streams. Remember that the lever-arm rule will work in any part of the diagram, whether it is two-phase or not (but extraction cannot perform a separation in the single-phase region). [Remember: The end points of the line correspond to the compositions of the streams, while the ratios of lengths can then be used to find their flowrates.]

Summary

1 The lever-arm rule is a *graphical alternative to solving a mass balance*.
2 The points on the line $\overline{OFV}$ correspond to the compositions of the various streams.
3 This method is useful for solving:
  (a) for a ratio of stream flowrates (V/O or any other);
  (b) for a stream composition (x_{Ai}, etc.).

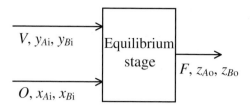

Figure 3.22 Equilibrium-limited stage for a three-phase system.

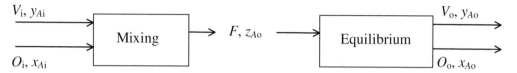

Figure 3.23 Mixing and separating stages in an extraction process.

4 To use the graphical method:
  - remember that there are two streams on one side of the mass balance and one stream on the other side ($V + O = F$);
  - note that the same method can be followed for one inlet stream which separates into two outlet streams (the reverse process).
    (a) Given the ratio V/O and the inlet compositions x_{Ai} and y_{Ai}, z_{Ao} could be located by solving the equation or measuring line segments:

$$\frac{V}{O} = \frac{x_{Ai} - z_{Ao}}{z_{Ao} - y_{Ai}} = \frac{\text{line segment } \overline{FO}}{\text{line segment } \overline{VF}}.$$

    (b) Measuring line segments would locate F, and then z_{Ao} could be found.
    (c) Given x_{Ai}, y_{Ai} and z_{Ao} (the inlet and outlet compositions), one could solve for V/O by measuring $\overline{VF}$ and $\overline{FO}$ and substituting into the above equation.

Special case: equilibrium stage

A special case of the lever-arm rule, which renders it applicable to extraction analysis, is an equilibrium-limited stage for a three-phase system (Figure 3.22). Everything stated for the lever-arm rule still applies here since the mass balances around the control volume (equilibrium stage) are still the same. The compositions of the three streams will still lie on a straight line, and stream ratios can still be calculated as before.

The second extraction step, separation of the two phases into the raffinate and extract streams, will be considered next. This step, when combined with the previous mixing analysis, constitutes one ideal equilibrium stage. Imagine (Figure 3.23) that a mixing is followed by an equilibrium stage which allows F to separate into two streams: V_o and O_o, which are in equilibrium.

Now $\overline{O_o F V_o}$ will also be a straight line based on a mass balance. Because the second stage is an equilibrium stage, $\overline{O_o F V_o}$ will be a tie line, which represents the equilibrium between y_{Ao} and x_{Ao}.

Table 3.3 *Equilibrium data (wt%).*

Furfural	Glycol	Water
94.8	0.0	5.2
84.4	11.4	4.1
63.1	29.7	7.2
49.4	41.6	9.0
40.6	47.5	11.9
33.8	50.1	16.1
23.2	52.9	23.9
20.1	50.6	29.4
10.2	32.2	57.6
9.2	28.1	62.2
7.9	0.0	92.1

Table 3.4 *Mutual equilibrium (tie-line) data (wt%) for furfural–ethylene glycol–water.*

Glycol in water layer	Glycol in solvent layer
7.7	28.9
6.1	21.9
4.8	14.3
2.3	7.3
11.5	41.8
32.1	48.8

Example 3.4 single equilibrium-limited stage (reproduced with permission of publisher, copyright © 1998, J. Wiley and Sons, Inc.)

Problem:

Ethylene glycol is mixed with water as anti-freeze in car engines. Rather than dispose of this mixture, removal of the ethylene glycol and recycle are preferred. Furfural (F) is suggested as a solvent for removing ethylene glycol (G) from water (W) in an extraction process. The equilibrium solubility and tie-line data for 25 °C are given in Tables 3.3 and 3.4 [10]. Using these data, construct an equilibrium phase diagram and:

(a) calculate the composition of the equilibrium phases produced when a 45 wt% glycol-in-water solution is contacted with its own weight in furfural. Show the process on the diagram.

(b) show the composition of the water–glycol mixture obtained if all the furfural is removed from the extract obtained in (a).

Solution:

(a) In Figure 3.24, the point F marks the initial solution: 0% F, 45% G, 55% W. As furfural is added to this mixture, the path from point F to point A is followed. When the point M has been reached, a mixture which is half furfural and half feed has been obtained (because $\overline{AM}/\overline{FM} = 1/1$). Notice that this is in the two-phase region. It will separate into two phases: the extract phase at point E (28.5% G, 6.5% W, 65.0% F), and the raffinate phase at point R (8% G, 84% W, 8% F). These points are found from the tie-line through point M. The compositions of the two phases can also be solved simultaneously:

$R + E = 200$ lb   (overall balance)

0.65(lb F/lb extract)$E + 0.08$(lb F/lb raffinate)$R = 100$ lb   (furfural balance)

$R = 53$ lb, and $E = 147$ lb.

Note that this answer could also have been obtained from the lever-arm rule:

$E = 200(\overline{RM}/\overline{RE})$ and $R = 200(\overline{EM}/\overline{ER})$.

(b) The path followed is from point E to point P on the diagram. The extract composition is 82.5% G, 17.5% W.

This problem can also be solved with the right triangle diagram in Figure 3.25. Note that the diagonal axis represents different solutions which are all 0% furfural. Even though the % furfural is difficult to pick off this graph because the horizontal and vertical axes are not the same length, remember that it can be found by subtracting the other two compositions from 100%.

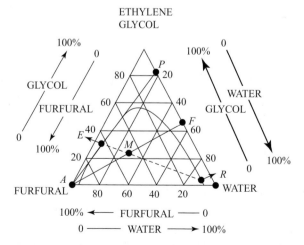

Figure 3.24 Furfural–ethylene glycol–water phase diagram [10]. Copyright © 1998, J. Wiley and Sons, Inc.

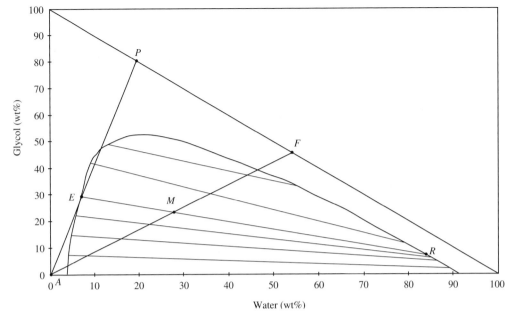

Figure 3.25 Furfural–ethylene glycol–water right triangle diagram.

Example 3.5

Problem:

Mixing oil, vinegar and spices for salad dressing (Figure 3.26).

While this is not an environmental application, this example is a common extraction situation which illustrates the concept.

Solution:

The mass balance equations could be solved algebraically if an algebraic expression for the phase equilibrium was known (an equation to relate y_{Ao} and x_{Ao}). Alternatively, the graphical method can be used to solve the mass balances and equilibrium relationships algebraically. Suppose the right triangular diagram for the ternary oil–vinegar–spices system is as shown in Figure 3.27. Here, point O_i represents the oil and spices, and point V_i represents the vinegar. Point F is where the mixing occurs (note that it is in the two-phase region). Point F separates into two phases along the tie-line (because they are in equilibrium) with point O_o representing the oil and spices layer and point V_o representing the vinegar and spices layer. If these layers were separated, an extraction would have been performed which used vinegar as the solvent to remove some of the spices (solute) from the oil (diluent).

Suppose one wanted to remove more spices by this same method. The oil and spices phase from the first equilibrium stage could be mixed with more vinegar. This is the idea behind a cross-flow cascade of equilibrium stages.

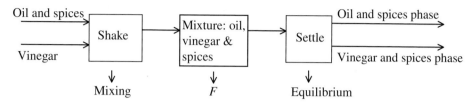

Figure 3.26 Salad dressing recipe, Example 3.5.

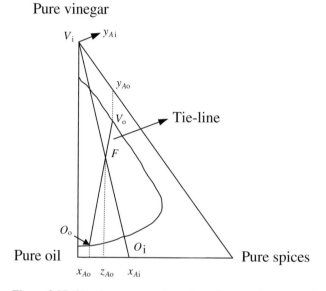

Figure 3.27 Oil, vinegar, and spices phase diagram, Example 3.5.

3.6   Binary feed mixtures

Referring to the generic separation schematic (Figure 3.28), assume that it represents a single stage of an equilibrium-based process for the separation of a binary feed mixture.

The assumption of equilibrium has two effects. First, the details of the flow and/or concentrations within the stage are not of interest. Second, streams L and V (exiting

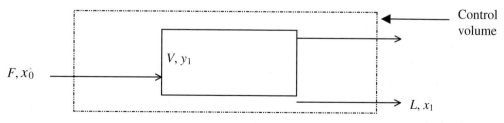

Figure 3.28 Generic binary feed separation stage for an equilibrium-based process.

streams) are in equilibrium with each other. The feed flowrate (F) and the mole fraction of our component of interest (x_0) are given. There are four unknowns, L, V, x_1, and y_1. Writing two independent mass balances and the equilibrium relation gives three equations:

Total mass balance: $\qquad\qquad F = L + V$ $\qquad\qquad\qquad$ (3.34)

Component mass balance: $\qquad x_0 F = y_1 V + x_1 L$ $\qquad\qquad$ (3.35)

Equilibrium relationship: $\qquad y_1 = mx$ (assume linear). $\qquad$ (3.36)

Therefore, a fourth relationship is needed. Typically, the ratio L/V can be set. Now, expressions for y_1 and x_1 can be derived in terms of the mass balances and equilibrium relationship (see if you can derive these)

$$y_1 = x_0 \frac{(1 + L/V)}{(1 + L/mV)} \qquad x_1 = \frac{x_0}{m} \frac{(1 + L/V)}{(1 + L/mV)}. \qquad (3.37)$$

As stated above, the analysis of a separation process uses mass balances in conjunction with some specific relation(s) which describe the separation process. For an equilibrium-stage process, this specific relation is the equilibrium relationship that describes the concentration of a component in each phase with respect to each other exiting the stage. Note that the equations can be solved for a single stage once the equilibrium relationship is known. It does not have to be a linear one.

Since the separation that is attainable is limited by the equilibrium between the two outlet streams, the next step would be to put several equilibrium stages in series. This is shown schematically in Figure 3.29.

The subscripts on x and y refer to the stage from which they *exit*. The quantities x_0 and y_{N+1} are inlet concentrations to the sequence of equilibrium stages. The flowrates L and V are assumed constant. This last assumption is useful for developing a basic understanding but is not a general requirement.

Two mass balances can be written (see Figure 3.29):

$$y_{N+1} = \frac{L}{V} x_N + \left(y_1 - \frac{L}{V} x_0 \right) \qquad (3.38)$$

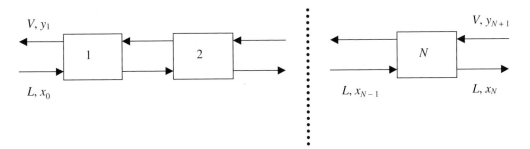

Figure 3.29 Equilibrium stages in series.

when our control volume contains N stages, and

$$y_{N+1} = \frac{L}{V}x_N + \left(y_N - \frac{L}{V}x_{N-1}\right) \tag{3.39}$$

for the Nth stage only.

Note that each balance describes the relationship between passing streams on each side of a stage (y_{N+1}, x_N) or sequence of stages. This is in contrast to the equilibrium relationship that describes two outlet streams from the same stage (y_N, x_N).

If the ratio L/V remains constant, then the above balances represent linear equations. Values for the concentrations on one end of the cascade would normally be known and the number of stages required for a given separation would be the variable of interest. As demonstrated above, a mass balance must be applied at each stage in combination with the equilibrium relationship to obtain the outlet concentrations for that stage. One method to perform this task is graphically. This method is referred to as the McCabe–Thiele method and the approach is commonly called "stepping off" stages. In Figure 3.30 the curved line is the equilibrium line and the straight line is our mass balance (operating line). Note that one is not limited to assuming a linear equilibrium relationship. The composition of one set of the passing streams for Stage 1 is (x_0, y_1). Starting at this point, one can move to the equilibrium line to obtain point (x_1, y_1), the composition of the exit streams from Stage 1. One then moves back to the operating line to point (x_1, y_2). This is the composition of the second set of passing streams for Stage 1. This "step" on the graph indicates one equilibrium stage in the cascade. One can continue in the same fashion and count the number of steps to obtain a given separation for the cascade.

The operating line can be above or below the equilibrium line but the stepping-off stages procedure remains the same. Start on the operating line with the passing streams on one side of the stage, go to the equilibrium line for the composition of the exit streams from that stage, then back to the operating line for the passing streams on the opposite side of the stage.

An equation can also be obtained to calculate the number of stages required. To simplify the analysis, some assumptions are made which reduce the complexity of the calculations and clarify the separation process:

(a) the flowrates of each stream are constant;

(b) the equilibrium relationship is a simple linear one (i.e., Henry's Law).

This situation corresponds to the transfer of a dilute solute between phases. Many environmental applications of separations involving extraction, distillation, and adsorption fall into this category so it is not a hypothetical example.

The normal method of cascading the equilibrium stages is using countercurrent flow. With two streams, they would enter the cascade at opposite ends of the cascade.

The most general form of a linear equilibrium relationship is

$$y = mx + b \tag{3.40}$$

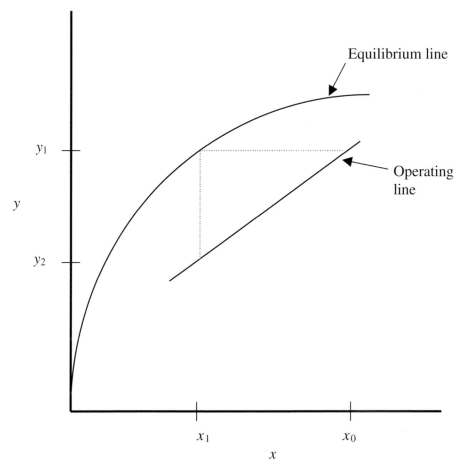

Figure 3.30 Graphical depiction of passing and exit streams for a single equilibrium-limited stage.

This form can be used to linearize a portion of a non-linear equilibrium relation, such as the one shown in Figure 3.30. By combining the mass balance with the equilibrium relationship, one can derive an equation that relates flowrates, and entrance and exit compositions to the number of equilibrium stages needed.

The mass balance around Stage 1 for the process shown in Figure 3.29 yields

$$y_2 = y_1 + \frac{L}{V}(x_1 - x_0).$$ (3.41)

Combining the equilibrium expression with this operating line gives

$$\frac{y_2 - y_1}{y_1 - mx_0 - b} = \frac{L}{mV}.$$ (3.42)

A similar balance around Stage 2 is

$$\frac{y_3 - y_2}{y_2 - y_1} = \frac{L}{mV}.$$ (3.43)

Multiplying the previous two equations gives

$$\frac{y_3 - y_2}{y_1 - mx_0 - b} = \left(\frac{L}{mV}\right)^2.$$ (3.44)

For any number P of equilibrium stages, a general equation can be formed

$$\frac{y_{P+1} - y_P}{y_1 - mx_0 - b} = \left(\frac{L}{mV}\right)^P.$$ (3.45)

To relate entrance and exit concentrations from the cascade, the previous equation for P equilibrium stages is summed over the entire cascade

$$\frac{y_{N+1} - y_1}{y_1 - mx_0 - b} = \sum_1^N \left(\frac{L}{mV}\right)^P.$$ (3.46)

Because y_1, the exit-phase concentration, is the unknown variable, we can rewrite this equation so that y_1 only appears in the numerator:

$$\frac{y_{N+1} - y_1}{y_{N+1} - mx_0 - b} = \frac{\sum_1^N \left(\frac{L}{mV}\right)^P}{1 + \sum_1^N \left(\frac{L}{mV}\right)^P}.$$ (3.47)

The summation of the power series when $L/mV < 1$ leads to

$$\frac{y_{N+1} - y_1}{y_{N+1} - mx_0 - b} = \frac{\frac{L}{mV} - \left(\frac{L}{mV}\right)^{N+1}}{1 - \left(\frac{L}{mV}\right)^{N+1}}.$$ (3.48)

Using the equilibrium relationship y_1^* to replace $mx_0 - b$

$$\frac{y_{N+1} - y_1}{y_{N+1} - y_1^*} = \frac{\left(\frac{L}{mV}\right) - \left(\frac{L}{mV}\right)^{N+1}}{1 - \left(\frac{L}{mV}\right)^{N+1}}.$$ (3.49)

This result is normally called the Kremser equation.

Note that, if $L/mV > 1$, then Equation (3.47) can be divided by $(L/mV)^N$ and the same result is obtained for Equation (3.49).

In this form, the Kremser equation is useful for solving problems where N is fixed and an exit composition needs to be calculated. When equilibrium is closely approached $(y_1 \rightarrow y_1^*)$, Equation (3.49) becomes

$$\frac{y_1 - y_1^*}{y_{N+1} - y_1^*} = \frac{1 - (L/mV)}{1 - (L/mV)^{N+1}}.$$ (3.50)

When the separation requirement is specified but N is unknown, the above equation can be rearranged to solve for N:

$$N = \frac{\ln\left\{[1 - (mV/L)]\left[(y_{N+1} - y_1^*)/(y_1 - y_1^*)\right] + (mV/L)\right\}}{\ln(L/mV)}. \tag{3.51}$$

The above equations are limited to cases of constant flowrates and linear equilibrium relationships. For situations where there are small deviations from linear phase equilibrium and/or changes in flow from stage to stage, the above equations can be applied over sections of the cascade in series. For situations where this approach is not reasonable, finite difference mathematical analysis can also be applied to equilibrium-stage calculations.

Example 3.6: gas stream with H$_2$S

Problem:

A waste-gas stream from a chemical processing facility contains 10% H_2S on a dry basis. The remainder is primarily N_2. This stream cannot be discharged directly into the environment. One approach to the removal of H_2S prior to discharge of the gas is to contact it with water in a staged column.

The target is 1% H_2S in the exit gas stream. The entering gas stream is saturated with water so there is no mass transfer of water to the gas phase. The solubility can be described by:

$$y_{H_2S} = 500 x_{H_2S}.$$

Initially, the ratio L/V is set to 750 to compensate for the low solubility of H_2S in H_2O (i.e., use more water so the total amount sorbed into the liquid phase is greater). Decide:

(a) how many equilibrium stages are needed;

(b) could this analysis be used if the entering concentration of H_2S was larger (50%, for example)?

Solution:

The equilibrium relationship is given. The slope of the operating line is also given. To use the McCabe–Thiele analysis, one point on the operating line is needed. The exit gas composition is 1% of H_2S ($y = 0.01$) and the entering water does not contain any H_2S ($x = 0$). Since the points represent passing streams on one end of the cascade, they represent one point on the operating line.

The McCabe–Thiele plot is drawn on Figure 3.31. Note that the scales are different on each axis, based on the equilibrium relationship and operating-line coordinates.

One can step off stages starting at either end of the cascade. From the plot, it can be seen that <u>four equilibrium stages</u> are needed.

The calculation can also be performed using the Kremser equation:

$$N = \frac{\ln\{[1 - (500/750)][0.1 - 0/0.01 - 0] + (500/700)\}}{\ln(750/500)} = 3.4.$$

Rounding up, it is again determined that <u>four equilibrium stages</u> are needed.
[Note that this analysis would be appropriate for the case of a more concentrated H_2S stream entering the column, as long as the equilibrium relationship still holds at that concentration.]

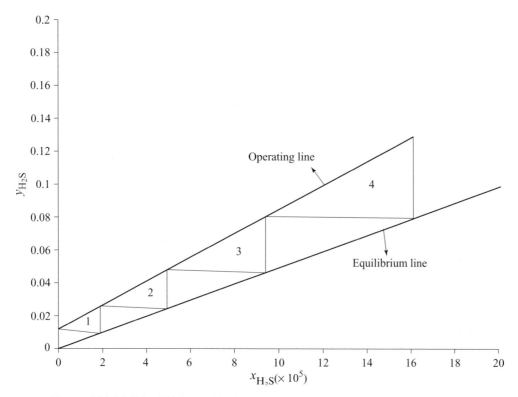

Figure 3.31 McCabe–Thiele graphical method for determining the number of equilibrium stages, Example 3.6.

3.7 Minimum number of stages

Another important concept is the calculation of the minimum number of equilibrium stages to perform a given separation. This is a theoretical concept since it occurs when one or both flowrates in a countercurrent cascade are infinite. It is useful to calculate since the actual number of stages will be larger (i.e., this is a limit). An analytical equation can be

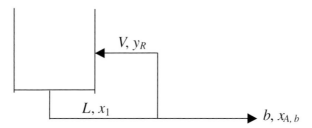

Figure 3.32 Bottom stage of a cascade.

derived for the cascade described above [11]. A constant value of the relative volatility, α_{AB} (see Equation (3.23)), is assumed for each stage.

Consider the bottom of any equilibrium-stage process as shown in Figure 3.32, where L is the liquid stream exiting the cascade, b is the product stream, and V is the stream which is vaporized and fed back into the cascade.

From the definition of a separation factor (Equation (2.9)), one can write

$$\left(\frac{y_A}{y_B}\right)_R = \alpha_R \left(\frac{x_A}{x_B}\right)_b,$$
(3.52)

where A and B denote the two components of a binary mixture. A mass balance yields

$$V y_R = L x_1 - b x_{A,b}.$$
(3.53)

The case of minimum number of equilibrium stages corresponds to infinite flow from and back into the cascade, such that no product is removed. In this case

$$V = L \quad \text{and} \quad b = 0.$$

The mass balance, then, reduces to

$$y_R = x_1.$$
(3.54)

Substitution back into the separation factor equation gives

$$\left(\frac{x_A}{x_B}\right)_1 = \alpha_R \left(\frac{x_A}{x_B}\right)_b.$$
(3.55)

To relate equilibrium Stage 2 to equilibrium Stage 1 and the final stage R, one can write

$$\left(\frac{x_A}{x_B}\right)_2 = \alpha_1 \left(\frac{x_A}{x_B}\right)_1 = \alpha_1 \alpha_R \left(\frac{x_A}{x_B}\right)_R.$$
(3.56)

This development can be followed to the top of the cascade for N equilibrium stages

$$\left(\frac{x_A}{x_B}\right)_N = \alpha_N \alpha_{N-1} \ldots \alpha_1 \alpha_R \left(\frac{x_A}{x_B}\right)_b,$$
(3.57)

or, assuming α is constant,

$$\left(\frac{x_A}{x_B}\right)_d = \alpha_{AB}^{N_{\min}} \left(\frac{x_A}{x_B}\right)_b , \tag{3.58}$$

where subscript d denotes the top of the cascade where component A is enriched and b denotes the bottom of the cascade where component B is enriched. This equation can be rearranged to solve for the minimum number of stages

$$N_{\min} = \frac{\ln\left[(x_A/x_b)_d / (x_A/x_B)_b\right]}{\ln \alpha_{AB}}. \tag{3.59}$$

This equation is called the Fenske–Underwood equation.

Note that the minimum number of equilibrium stages increases non-linearly as the separation required becomes more difficult [more extreme x_A/x_B ratio or α_{AB} closer to unity].

Equation (3.59) illustrates two important points: (1) the value of the minimum number of stages is independent of the feed conditions and only depends on the separation requirements; and (2) increasing the number of stages is usually more effective than increasing flow to increase product purity for difficult separations.

Example 3.7

Problem:
Using the information in the previous example, calculate $N_{\min}$. The binary system is H_2S and N_2. The water is essentially an immiscible liquid since it is in low concentration in the gas phase and there is no net mass transfer of water.

$$y_{N_2} = 8.6 \times 10^4 x_{N_2}.$$

Solution:
For equilibration of immiscible phases, one can write

$$\alpha_{AB} = \frac{m_A}{m_B}.$$

Since N_2 is the component being enriched,

$$\alpha_{AB} = \frac{8.6 \times 10^4}{5.0 \times 10^2} = 172,$$

and

$$N_{\min} = \frac{\ln\left[(99/1)/(90/10)\right]}{\ln(172)} = 0.47.$$

Scaling up, one equilibrium stage would work as $L/V \rightarrow$ infinity. This result emphasizes the point that it is better to increase the number of stages than increase the volumes (or flowrates).

3.7.1 Allowance for stage efficiencies

There are two common approaches to include the influence of non-equilibrium on the quality of separation achieved for a given number of stages or on the number of actual stages required for a given separation requirement. The first of these involves the use of an *overall efficiency* E_o defined as

$$E_o = \frac{\text{number of equilibrium stages}}{\text{number of actual stages}}. \tag{3.60}$$

In order to use the overall efficiency in a design problem, one carries out an equilibrium-stage analysis and then determines the number of actual stages as the number of equilibrium stages divided by E_o. Thus the overall efficiency concept is simple to use once E_o is known, but it is often not easy to predict reliable values of E_o.

The other commonly used approach involves the concept of the *Murphree vapor efficiency* E_{MV} defined as:

$$E_{MV} = \frac{y_1 - y_{N+1}}{y_N^* - y_{N+1}} \tag{3.61}$$

where y_N^* is the vapor composition which would be in equilibrium with the actual value of x_N. There is more theoretical basis for correlating and predicting values of E_{MV} than is the case for E_o.

For the case of a constant Murphree efficiency, the Kremser equation, Equation (3.49), can be modified to calculate the actual number of stages:

$$N = \frac{\ln\left\{[1 - (mV/L)]\left[\left(y_{N+1} - y_1^*\right)/\left(y_1 - y_1^*\right)\right] + (mv/L)\right\}}{\ln\left\{1 + E_{MV}\left[(mV/L) - 1\right]\right\}}. \tag{3.62}$$

If the value of E_{MV} is known for each stage in a binary separation (or is taken at a single known constant value for the whole sequence of stages), it may be readily used in the McCabe–Thiele graphical construction. Chapter 4, on distillation, has a section which illustrates this approach.

3.8 Rate-limited processes

Analysis of rate-limited processes usually begins with a *differential mass balance*. One description of this balance for our stationary control volume for a component A is:

$$\text{In} + \text{Generation} = \text{Out} + \text{Accumulation}. \tag{3.63}$$

Rearranging,

$$\text{Accumulation} = (\text{In} - \text{Out}) + \text{Generation}. \tag{3.64}$$

In mathematical notation, this is written as:

$$\frac{\partial C_A}{\partial t} = -\nabla \cdot N_A + R_A, \tag{3.65}$$

where C_A = molar concentration of A

N_A = total molar flux of A (convection + diffusion)

R_A = homogeneous reactions involving A within the control volume

∇ = del operator.

The term R_A does not account for reactions at the boundary of the control volume (typically a surface where a heterogeneous reaction occurs). Reactions at boundaries are accounted for by boundary conditions when solving the differential balance.

If the system is not changing with respect to time, then $\partial C_A / \partial t = 0$. If no reactions involving A are taking place within the control volume, then $R_A = 0$. Equation (3.33) reduces to

$$\nabla \cdot N_A = 0. \tag{3.66}$$

If the physical meaning of this term is evaluated, in = out. For one-dimensional planar transport (i.e., x direction)

$$\frac{dN_A}{dz} = 0 \Rightarrow N_A = \text{constant.} \tag{3.67}$$

This result indicates that the mass flux of A (and the rate since the cross-sectional area is constant) remains constant as one moves in the x direction). For cylindrical systems (tubes) where radial transport occurs,

$$\frac{d(rN_A)}{dr} = 0 \Rightarrow rN_A = \text{constant} \Rightarrow N_A = \frac{\text{constant}}{r}. \tag{3.68}$$

This result demonstrates that the flux is not constant but varies as r^{-1}. The cross-sectional area is $2\pi rL$ (where L is axial length) so the rate is a constant. Prove this for yourself.

An equation is needed for N_A to substitute into the above mass balance. For a binary system (A and B), the molar average velocity of flow (v_M) for both components is

$$v_M = \frac{N_A + N_B}{C}; \qquad C = \text{total molar concentration.} \tag{3.69}$$

The velocity for component A is

$$v_A = \frac{N_A}{C_A}. \tag{3.70}$$

Note that $x_A = C_A / C$.

The equation for v_M can be rewritten as

$$v_M = \frac{N_A}{C} + \frac{N_B}{C} = \frac{C_A}{C} \frac{N_A}{C_A} + \frac{C_B}{C} \frac{N_B}{C_B} = x_A v_A + x_B v_B. \tag{3.71}$$

The velocity for component A has two contributions, the molar average velocity of the system, v_M, plus the movement due to diffusion of A, v_{AD}:

$$v_A = v_M + v_{AD}. \tag{3.72}$$

It can be readily observed that $v_A = v_M$ if the convective flow is dominant and $v_A = v_{AD}$ if there is no convection. An equation can be written for v_{AD} analogous to the equation for v_A:

$$v_{AD} = \frac{J_A}{C_A}; \qquad J_A = \text{molar diffusion flux.} \tag{3.73}$$

Returning to the definition of v_A:

$$v_A = \frac{N_A}{C_A} \Rightarrow N_A = C_A v_A \tag{3.74}$$

$$N_A = C_A (v_M + v_{AD}) = \frac{C_A}{C} C v_M + C_A v_{AD} \tag{3.75}$$

$$= x_A (N_A + N_B) + J_A. \tag{3.76}$$

For one-dimensional planar transport, substituting Fick's Law for J_A yields

$$N_A = x_A (N_A + N_B) - D_{AB} \frac{dC_A}{dz}. \tag{3.77}$$

To simplify this equation, ask the following:
1 Is N_A constant (from mass balance)?
2 Is C a constant? $dC_A/dz = C\,(dx_A/dz)$?
3 What is flux ratio (N_A/N_B)? For example, is $N_B = 0$?
4 Is $x_A \ll 1$? If so, the first term on the right can be neglected.

The mass balance equation plus the equation for N_A combine to produce a differential equation. To solve it, boundary conditions are needed.

Concentrations are specified
  (a) At an interface (such as phase equilibrium with surrounding fluid);
  (b) for an instantaneous irreversible reaction at boundary ($C_A = 0$);
  (c) at infinity if the mass transfer does not reach the opposite boundary ($C_A = $ initial concentration).

Flux is specified
  (a) At an interface;
  (b) at a solid surface;
  (c) as a heterogeneous reaction at a surface.

There are references [12, 13] that discuss diffusional processes in detail.

An alternative analysis to a differential mass balance is based on *mass transfer coefficients*. This analysis is useful when detailed analysis of the concentration in the system is not needed. This approach can be used in terms of resistances to mass transfer.

There can often be more than one resistance to mass transfer. These resistances can include boundary layer and diffusion effects. For example, in adsorption a solute must diffuse through a fluid, cross the boundary layer between the fluid and the solid sorbent, then diffuse into the sorbent particle. The configuration of a process can also affect the overall mass transfer resistance. Therefore, the flux, or rate per unit area, of a component A is usually expressed in words as:

$$\text{Flux of } A = (\text{Overall Mass Transfer Coefficient}) \times (\text{Driving Force}). \qquad (3.78)$$

This equation can be rearranged as:

$$\text{Flux of } A = \frac{\text{Driving Force}}{\text{Total Mass Transfer Resistance}} = \text{OMTC} \times \text{Driving Force}, \qquad (3.79)$$

where the OMTC includes all system contributions to mass transfer resistance.

The value of the total mass transfer resistance is the inverse of the overall mass transfer coefficient value. This equation is analogous to Ohm's Law which relates current flow (flux) to applied voltage (driving force):

$$i = \frac{\Delta V}{R}. \qquad (3.80)$$

There are different approaches that can be taken to estimate the OMTC (or resistance). The first is to directly measure the flux and driving force and calculate the coefficient. The second uses correlations that are available to estimate the value based on the particular process and operating conditions. Examples of this approach will be included in later chapters that deal with a particular separation technology. The third is to determine each mass transfer resistance and combine the terms to calculate the total resistance. This approach is analogous to the calculation of an equivalent resistance for an electrical circuit.

The two-film model is a simple example of this approach. A system of two fluids exists, with a distinct interface between the two (gas/liquid or two immiscible liquids). For purposes of this example, we assume a gas/liquid interface; Figure 3.33 illustrates the region near the interface. There will be a film (or boundary layer) on each side of the interface where, due to mass transfer from one phase to the second (gas to liquid in the figure), the concentration of A is changing from its value in the bulk phase, $P_{A,b}$ in gas and $C_{A,b}$ in liquid. The thickness of the boundary layer on each side of the interface will typically be different and a function of the fluid and flow conditions in each phase. At steady-state, the flux of A can be described as:

$$J_A = \underbrace{k_g(P_{A,b} - P_{A,i})}_{\text{Gas Phase}} = \underbrace{k_L(C_{A,i} - C_{A,b})}_{\text{Liquid Phase}}, \qquad (3.81)$$

where k_g and k_L are the mass transfer coefficients in each phase. These values can be estimated from correlations based on flow conditions and configuration. An equation can also be written for the equilibrium at the interface (such as Henry's Law):

$$P_{A,i} = m C_{A,i}. \qquad (3.82)$$

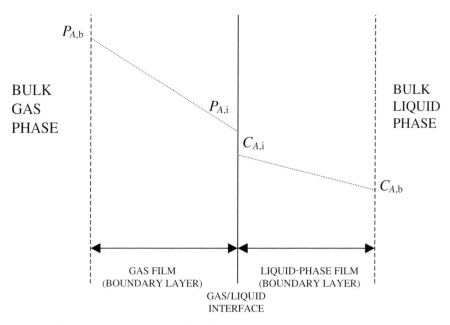

Figure 3.33 Region near a gas/liquid interface.

Using the above two equations, an equation can be obtained for the flux of A in terms of a driving force based on the bulk-phase concentrations (see Problem 3.2 at end of chapter):

$$J_A = \frac{1}{\left(\dfrac{1}{k_g} + \dfrac{m}{k_L}\right)}\left(P_{A,b} - mC_{A,b}\right). \tag{3.83}$$

There are some important points to note. As written, the driving force is in terms of the gas phase since $mC_{A,b}$ is the pressure that would be in equilibrium with the liquid-phase bulk concentration. An equation for J_A could also be derived in terms of liquid-phase concentration. The overall mass transfer resistance (R_T) has two contributions:

$$R_T = \underbrace{\frac{1}{k_g}}_{\text{Gas-Phase Resistance}} + \underbrace{\frac{m}{k_L}}_{\text{Liquid-Phase Resistance}}. \tag{3.84}$$

Some implications of this resistance term can be observed immediately. First, the larger the value of k, the smaller the resistance. Second, the value of each resistance can be different. When one resistance is significantly larger than the other (or the mass transfer coefficient for one phase is significantly smaller), it is dominant and is termed the controlling resistance. Mass transfer across both films is controlled (limited) by the dominant resistance. Third, the larger the value of m, the larger the liquid-phase resistance (can you see why physically?).

The flux equation can be written as:

$$J_A = K_G(P_{A,b} - P_A^*), \tag{3.85}$$

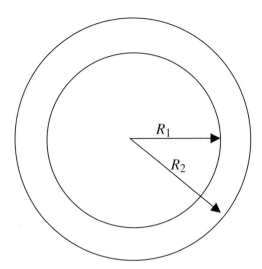

Figure 3.34 Tubular membrane.

where

$$\frac{1}{K_G} = \frac{1}{k_g} + \frac{m}{k_L}, \tag{3.86}$$

and

$$P_A^* = mC_{A,b}. \tag{3.87}$$

Here, K_G is an overall mass transfer coefficient. The equation for K_G is analogous to the resistance in series model for electrical circuits. P_A^* is the value of the solute pressure that would be in equilibrium with $C_{A,b}$.

An analogous equation can be written in terms of liquid-phase concentrations:

$$J_A = K_L(C_{A,b}^* - C_{A,b}), \tag{3.88}$$

where

$$P_{A,b} = mC_{A,b}^*. \tag{3.89}$$

More complicated mass transfer systems may not be amenable to determination of each mass transfer resistance. The flux is written in terms of an overall coefficient.

One type of system to consider is one in which the surface area perpendicular (normal) to the solute flux changes within the system, then the steady-state flux is <u>not</u> constant but the mass transfer rate is. As an example, consider Figure 3.34, in which the material between R_1 and R_2 is a membrane. At steady-state, mass can permeate from R_1 to R_2. The mass transfer rate R_A (mass/time) is constant. The flux J_A (mass/area · time) is related to the rate by:

$$R_A = J_A \times \text{(surface area normal to the mass transfer direction)}, \tag{3.90}$$

since $A_1 \neq A_2$, $J_{A1} \neq J_{A2}$.

For this type of system, the flux needs to be specified at a location.

Example 3.8

An air-stripping tower is to be designed to remove volatile organic compounds (VOCs) from water droplets. The overall mass transfer coefficient in the liquid phase is given by:

$$\frac{1}{K_L} = \frac{1}{k_L} + \frac{1}{mk_g}.$$

(3.91)

The distribution coefficient, m, is 1.8 atm $\cdot$ m^3/mg and the partial pressure of VOCs is 0.14 atm. Assume clean air is used such that $C_{A,b} = 0$. For a gas-phase mass transfer coefficient of 1.3×10^{-9} mg/atm $\cdot$ cm^2 $\cdot$ s, examine three designs for which the liquid-phase mass transfer coefficient is $6.2 \times 10^{-3}, 6.2 \times 10^{-2}$, and 6.2×10^{-4} cm/s, respectively.

Solution:

For the liquid phase:

$$J_A = K_L \left(C_{A,b}^* - C_{A,b} \right)$$

and

$$P_{A,b} = m C_{A,b}^*.$$

So,

$$C_{A,b}^* = \frac{0.14 \, \text{atm}}{1.8 \, \text{atm} \cdot \text{m}^3/\text{mg}} = 0.078 \, \text{mg/cm}^3$$

and

$$mk_g = \left(1.8 \, \frac{\text{atm} \cdot \text{m}^3}{\text{mg}} \right) \left(\frac{10^6 \, \text{cm}^3}{\text{m}^3} \right) \left(1.3 \times 10^{-9} \, \frac{\text{mg}}{\text{atm} \cdot \text{cm}^2 \cdot \text{s}} \right)$$
$$= 2.3 \times 10^{-3} \, \text{cm/s}.$$

Case I: $k_L = 6.2 \times 10^{-3}$ cm/s

$$\frac{1}{K_L} = \frac{1}{0.0062} + \frac{1}{0.0023} \qquad K_L = 1.7 \times 10^{-3} \, \text{cm/s}$$
$$J_A = 0.0017 \, \text{cm/s} \, (0.078 \, \text{mg/cm}^3 - 0) = 1.33 \times 10^{-4} \, \text{mg/cm}^2 \cdot \text{s}.$$

Case II: $k_L = 6.2 \times 10^{-2}$ cm/s

$$\frac{1}{K_L} = \frac{1}{0.062} + \frac{1}{0.0023} \qquad K_L = 2.26 \times 10^{-3} \, \text{cm/s}$$
$$J_A = 2.26 \times 10^{-3} \, \text{cm/s} \, (0.078 \, \text{mg/cm}^3 - 0) = 1.76 \times 10^{-4} \, \text{mg/cm}^2 \cdot \text{s}.$$

Case III: $k_L = 6.2 \times 10^{-4}$ cm/s

$$K_L = 4.9 \times 10^{-4} \, \text{cm/s}$$
$$J_A = 0.00049 \, \text{cm/s} \, (0.078 \, \text{mg/cm}^3 - 0) = 3.82 \times 10^{-5} \, \text{mg/cm}^2 \cdot \text{s}.$$

Hence, as k_L increases, so does the rate of mass transfer.

In addition, for Case II, k_G is the limiting rate and hence $K_L \approx k_g$. For Case III, k_L is the limiting rate and $K_L \approx k_L$.

3.9 Remember

- Mass balances can be written for both the total mass in a system and a specific component of the system.
- Equilibrium processes are limited by the degree of separation in a single contact, so cascades of contact stages are used to achieve a higher degree of separation.
- The minimum number of equilibrium stages is a theoretical concept based upon infinite flows within a cascade and no feed or product streams.
- Equilibrium analysis of separation processes can be performed both graphically and numerically by the use of operating lines (obtained from mass balances) and equilibrium lines.
- Mass transfer (rate) analysis of a process is characterized by an overall mass transfer coefficient (OMTC), which is a global term that includes system contributions to mass transfer resistance.

3.10 Questions

3.1 Define a microscopic and macroscopic balance.

3.2 State the basis for an equilibrium- and rate-limited process.

3.3 Describe the assumptions about solute–sorbent interactions for the following isotherms:

Langmuir-single component

Langmuir-competitive

Freundlich.

3.11 Problems

3.1 Use liquid-phase rather than gas-phase compositions and derive the equation below. For this case, the flows are in a direction opposite to those in Figure 3.6.

$$N = \frac{\ln\left\{[1 - (L/mV)]\left[(x_{N+1} - x_1^*)/(x_1 - x_1^*)\right] + (L/mV)\right\}}{\ln(mV/L)}.$$

3.2 Derive an equation for J_A analogous to Equation (3.83) in terms of liquid-phase concentrations.

3.3 For vapor–liquid equilibrium, the ideal separation factor is termed the relative volatility and is written as the ratio of the vapor pressure of each component:

$$\alpha_{ij} = \frac{P_{v_i}}{P_{v_j}}.$$

For <u>constant</u> relative volatility, derive the equation:

$$y_i = \frac{\alpha_{ij} x_i}{1 + (\alpha_{ij} - 1)x_i}.$$

3.4 Ammonia contained in an air stream needs to be removed prior to discharge. One method is to sorb the ammonia using water in a countercurrent column. The inlet water stream contains 0.1 mole% ammonia. The outlet concentration must be lower than $y_{NH_3} = 1 \times 10^{-5} = 1 \times 10^{-5}$ (mole fraction). For this dilute stream, $y = 1.4x$. It is proposed to use an L/V ratio equal to 2.8.

(a) How many equilibrium stages are needed?

(b) How much would the air flowrate (or L/V ratio) be changed to decrease the answer in part (a) by one equilibrium stage?

3.5 A composite film is made up of two regions I and II (Figure 3.35) [14]. [Copyright © 1998, John Wiley & Sons, Inc. This material is used by permission of John Wiley & Sons, Inc.]

This composite film is used to remove CO_2 from a gas stream. The partition coefficient and diffusion coefficient of CO_2 are different in each region. Assume that the

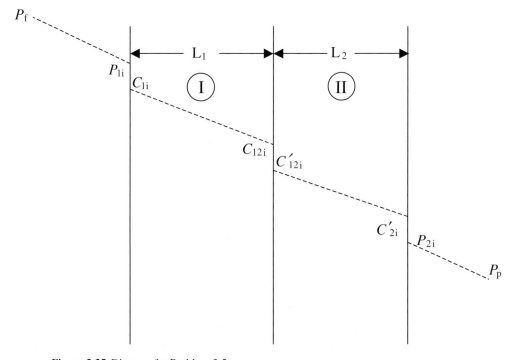

Figure 3.35 Diagram for Problem 3.5.

gas-phase mass transfer coefficient (k_g) on each side of the composite film is the same.

For steady-state mass transfer, does it make any difference which side of the film is the feed phase? Prove your answer and state why. The diagram has feed on the left. Assume P_f and P_p are constant.

$$P_1 = m_1 C_1$$
$$C_{12} = m_2 C'_{12}$$
$$P_2 = m_3 C'_2$$

3.6 An organic extraction from water into various solvents is being considered. Some solvents and their distribution coefficients (m) are:

Solvent	A	B	C
m	3	30	90

The distribution coefficient is based on mass concentrations in each phase.

(a) Determine the number of equilibrium stages needed to reduce the organic concentration from 1000 ppm to 1 ppm for $mV/L = 1$ and 2 ($V =$ solvent mass flowrate and $L =$ aqueous feed mass flowrate).

(b) For $mV/L = 1$, determine V for each solvent for $L = 10$ kg/min.

3.7 Irrigation run-off exits three different farms and merges into a single stream. The average flowrates, phosphate levels and nitrogen levels in each run-off stream are shown in the table. Perform an overall and component mass balance to obtain the total flowrate, phosphate and nitrogen levels of the merged stream.

	Flowrate (L/min)	Phosphate (mg/L)	Nitrogen (mg/L)
Stream 1	1200	30	25
Stream 2	1450	15	35
Stream 3	1300	25	30

3.8 A distillation column separates a feed stream of 50 wt% methanol and 50 wt% water. The feed enters at a rate of 100 kgmol/hr. The overhead product can contain no more than 1 mol% water and the bottoms product can contain no more than 4 mol% methanol. What are the rates of the overhead and bottom streams?

3.9 A ternary system of 60 wt% methanol, 10 wt% butanol, and 30 wt% water is the feed stream for an alcohol/water separation unit. There are three exit streams. The composition of one exit stream is 95 wt% methanol, 4.5 wt% butanol, and 0.5 wt% water. The second stream has 96 wt% butanol, 3 wt% methanol and 1 wt% water. The third stream is 95 wt% water with 2.5 wt% each of methanol and butanol. Calculate the total mass flowrate of the three exit streams per kg feed.

3.10 Storm sewer drains from three subdivisions feed into a single canal that flows to a municipal water treatment facility. Each subdivision pays the municipal facility based upon the total quantity of water and the concentration of organics in the water.

The monitoring devices for the first subdivision have been rendered inoperable. Given an overall canal flow of 6400 gpm (gallons per minute) with an organics load of 240 mg/L, estimate the flow and organics load in the first sewer if the other two sewers are measured at 1850 gpm with 280 mg/L organics and 2250 gpm with 210 mg/L organics.

3.11 A distillation column is needed to separate methanol and water. The overhead stream should be at least 97% methanol and the bottom stream should contain no more than 3% methanol. Apply the graphical method to estimate the minimum number of stages to achieve this separation. Assume the equilibrium relationship, $y^* = 1.3x$, is linear and the operating line is the diagonal.

3.12 An absorption column is built to remove CO_2 from an air stream leaving a combustion process. The entering gas stream is 10 mol% CO_2 and the exit gas stream must be no greater than 0.5 mol%. The entering liquid stream contains 0.1 mol% CO_2. The equilibrium relationship, $y^* = 2.3x$, is linear and the L/V ratio is 4.

(a) Perform a mass balance to obtain the CO_2 concentration in the exit liquid stream.

(b) Apply the graphical method to estimate the number of stages. If the overall stage efficiency is 0.65, how many stages are required?

(c) Estimate the number of steps using the Kremser equation.

3.13 Evaluate the number of required stages for Problem 3.12 for an L/V ratio of 3 and 5. Assume the entering and exiting gas-phase concentrations do not change. How does the L/V ratio affect the number of stages required?

3.14 For the absorption column in Problem 3.12, estimate the number of equilibrium stages required for equilibrium relationships of $y^* = 1.2x$ and $y^* = 3.8x$, respectively. How does increasing the Henry's Law relationship affect the number of required stages?

3.15 A distillation column is designed to separate benzene and toluene. The relative volatility ratio for this binary system is 2.5. If the overhead stream must be at least 97% benzene and the bottoms stream no greater than 6% benzene, apply the Fenske–Underwood equation to estimate the minimum number of stages required.

3.16 For the separation defined in Problem 3.15, estimate the minimum required stages under the following conditions:

(a) benzene in overhead = 99.5%, benzene in bottoms = 6%;

(b) benzene in overhead = 92%, benzene in bottoms = 6%;

(c) benzene in overhead = 99.5%, benzene in bottoms = 0.5%.

What effect do both the overhead and bottoms product purity have on the required number of stages?

3.17 Experiments are performed to characterize the flux of carbon dioxide through a membrane. If the flux is measured at 2.09×10^{-11} mol/cm$^2 \cdot$ s, the concentration on the incoming side of the membrane is 0.5 mg/L, and the concentration on the opposite side of the membrane is 0.1 mg/L, what is the overall mass transfer coefficient?

3.18 A solid carbon adsorbent is to be used to remove 53 mg/L chlorides from a water stream. There are two mass transfer resistances. The first, resistance to the movement of chlorine through the water to the surface of the adsorbent particle, is 0.0017 cm/s. The second, resistance to movement of chlorine into the solid particle, is 0.000 65 cm/s. If all of the chlorine is to be removed and the solid adsorbent initially has no chlorine, what is the maximum flux of chlorine in $mg/cm^2 \cdot s$?

3.19 A solution of 48% sucrose (sugar) by weight contains a small quantity of impurities. The impurities will be removed with a carbon adsorbent. The equilibrium follows a Freundlich isotherm, $Y = mX^n$, where Y has units of mg impurity/kg sugar and X has units of mg impurity/kg carbon. The isotherm constants are $n = 2.2$ and $m = 8.0 \times 10^{-7}$.

    (a) For a single-stage process, how much carbon per 1000 kg of feed solution is needed to reduce the impurity level from 20 mg to 2.5% of its original value?

    (b) Consider a countercurrent operation in which 5 kg of carbon is contacted with 1000 kg of the feed solution. How many stages are needed to reduce the impurity level from 20 mg to 1 mg?

3.20 Adsorption of organics using activated carbon follows a linear isotherm: $(x/m)KC_e$. To process a volume V of fluid containing an initial concentration C_0 of organics, it is suggested to divide the total mass M of sorbent into three equal parts and contact the fluid with $M/3$ mass of sorbent in three consecutive batch contacts, allowing equilibrium to be reached with each contact.

    (a) Find the concentration after each batch contact in terms of initial concentration C_0 and constants.

    (b) Find the total amount of organics sorbed in terms of C_0 and constants.

    (c) Explain the advantages and disadvantages of this approach compared to a single contact with mass M of sorbent.

4

Distillation

Double, double, toil and trouble, fire burn, and cauldron bubble.

– WILLIAM SHAKESPEARE, *Macbeth*

4.1 Objectives

1 Describe the physical meaning of:
  (a) constant molar overflow (CMO);
  (b) an operating line;
  (c) an equilibrium line.
2 For the McCabe–Thiele graphical method:
  (a) Draw an operating line for each section of a distillation column.
  (b) Determine minimum and actual reflux ratios, and know the meaning of total reflux.
  (c) Handle multiple feeds and/or sidestreams.
  (d) Determine the number of equilibrium stages.
  (e) Given the Murphree efficiency, determine the actual number of stages required.
  (f) Draw the q-line (which is a measure of the vapor and liquid content) for the feed stream in a McCabe–Thiele diagram.
  (g) Locate optimum feed plate location.
  (h) Step off stages when feed is not located on optimum plate.
  (i) Determine the number of equilibrium stages when using steam injection instead of a reboiler.

4.2 Background

Distillation is the most widely used separation method in the chemical and petrochemical industries. It is not usually considered initially for environmental applications but can serve as a useful approach. It is, however, frequently utilized to recover the mass-separating agent for recycle back into the system following absorption, stripping and extraction processes. Some environmental applications of distillation are the separation of organic solvent/water mixtures and water removal for volume reduction prior to disposal of hazardous waste mixtures. In addition, the principles used to design and analyze distillation columns are also applicable to other separation methods (extraction, absorption, stripping). So, the knowledge gained from the subject matter in this chapter will be applied in the study of these other methods.

Distillation is an equilibrium-limited separation which uses heat as an energy-separating agent. It is applied when two or more relatively volatile liquids, that vaporize at different temperatures, need to be separated or fractionated into almost pure product streams. Distillation separates components of a liquid mixture based on their different boiling points. When the boiling points of the entering species are significantly different, distillation can easily separate the feed into almost pure product streams of each component. However, as the boiling points become closer, distillation requires a large number of equilibrium stages to perform the separation.

Distillation is the baseline process for the chemical process industry, with 40,000 columns in operation in the US, handling 90–95% of all separations for product recovery and purification. The capital invested in distillation systems in the US alone is at least $8 billion [1].

Because it is prevalent throughout industry, there are numerous advantages to distillation as a separation technology. The process flow sheets are relatively simple and no mass-separating agent is required. The capital costs are low as is the risk associated with lesser known technologies. There is an abundance of data describing vapor–liquid equilibrium for many systems. Usually, distillation can be designed using only physical properties and vapor–liquid equilibrium (VLE) data, so scale-up is often very reliable. The primary disadvantage of distillation is its lack of energy efficiency. It is not a useful technique for systems containing an azeotrope or those with close boiling points. An azeotrope occurs when the vapor and liquid compositions of a system become identical such that, without altering the system, no further separation is possible. Distillation also cannot be applied to feed streams which are sensitive to thermal degradation or that polymerize at elevated temperatures. Operating the column under vacuum, however, can reduce or eliminate these problems.

The separation in a distillation process is governed by a difference in the composition of a liquid and vapor phase. This difference is usually characterized by a difference in actual vapor pressures, or volatilities, of the liquid-phase components. Vapor–liquid equilibrium data for the mixture components are, therefore, an important element for design and analysis. For a binary system we can define the relative volatility of the two species to be separated in terms of the more volatile component, A, with respect to the less volatile one, B:

$$\alpha_{AB} = \frac{K_A}{K_B} = \frac{y_A/x_A}{y_B/x_B} = \frac{y_A/x_A}{(1 - y_A)/(1 - x_A)}. \tag{4.1}$$

Higher relative volatilities correspond to easier separation by distillation. The greater the slope of the vapor vs liquid (y_A vs x_A) equilibrium line, or the larger the difference between y_A and x_A, the easier the separation. The relative volatility must exceed one but, realistically, must be greater than 1.3 for distillation processes [2]. It is important to note that α_{AB} is dependent on pressure and that it is not necessarily constant over a composition range.

4.3 Batch distillation

Batch distillation is often applied to separations in which small amounts of materials are processed or in which the plant does not operate continuously. It is versatile, such that the same equipment can be used for several products of varying composition at different times. The distillate, which is typically much more volatile than the liquid from which it is being recovered, is usually the product.

In batch distillation, a feed mixture is charged into a batch still where steam heat is supplied. The product is removed from the top of the column until the process is complete, after which the steam supply is discontinued and the remaining liquid in the still is removed. Batch systems, as shown in Figure 4.1(a) and (b), can either be single staged, in which the feed is charged into the reboiler, or multistaged, where a packed or tray column is placed above the reboiler. Single-staged units are similar to flash distillation units, except that feed and product streams are not continuous. In multistaged processes, reflux is returned to the column during the duration of the run. Product can either be removed continuously during the run or stored in the accumulator until the composition is of the desired quality. In the unusual case where the bottoms is the desired product, inverted batch distillation is applied. In this case, the bottoms liquid is continuously removed.

Because it is not continuous, the mathematical analysis of batch distillation is based on the total quantities. For a binary system in which the distillate is the desired product, the overall mass balance at the end of a batch run is

$$F = B + D \tag{4.2}$$

where F is the mass of feed charged into the process initially and D and B are the final masses of distillate and bottoms, respectively. For the desired component, the mass balance is

$$F x_F = B x_{B,f} + D x_{D,avg} \tag{4.3}$$

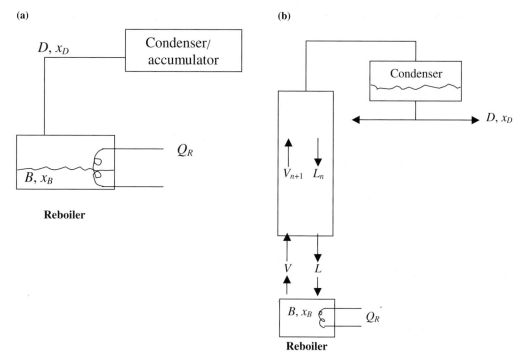

Figure 4.1 (a) Single-staged, and (b) multistaged batch distillation processes.

where x_F is the feed mass fraction, $x_{B,f}$ is the final mass fraction in the bottoms, and $x_{D,\text{avg}}$ is the average mass fraction drawn from the distillate. Both F and x_F are known quantities and normally either $x_{B,f}$ or $x_{D,\text{avg}}$ is specified. The above two balances combined with the Raleigh equation are used to solve for the three unknowns, D, B and either $x_{B,f}$ or $x_{D,\text{avg}}$. The Raleigh equation, presented below, is a differential mass balance that assumes the hold-up in the accumulator and column is negligible, such that only hold-up in the bottom stage or reboiler is significant. Then, if a differential quantity dB of composition x_D is removed from the system, the mass balance becomes:

$$-x_D dB = -d(Bx_B).$$ (4.4)

Expanding and solving the Raleigh equation yields

$$\ln\left(\frac{B}{F}\right) = -\int_{x_{B,f}}^{x_F} \frac{dx_B}{x_D - x_B}.$$ (4.5)

Performing this integration requires that x_D be related to x_B. In single-stage distillation cases where constant relative volatility can be assumed, the equilibrium relationship is:

$$y = \frac{\alpha x}{1 + (\alpha - 1)x},$$ (4.6)

and the Raleigh equation can be integrated analytically to obtain:

$$\ln\left(\frac{B}{F}\right) = \frac{1}{\alpha - 1} \ln\left[\frac{x_{B,f}(1 - x_F)}{x_F(1 - x_{B,f})}\right] + \ln\left[\frac{1 - x_F}{1 - x_{B,f}}\right]. \tag{4.7}$$

When constant relative volatility cannot be assumed, either graphical or numerical techniques, such as Simpson's rule, must be applied [2].

4.3.1 Multistaged batch distillation

In multistaged systems, x_D and x_B are not in equilibrium and, hence, integration of Equation (4.5) requires a different relationship between these product compositions. This is obtained by stage-by-stage calculations. If one assumes negligible hold-up at each stage and at the condenser and accumulator, mass balances can be written for any time t during the batch operation,

$$V_{n+1} = L_n + D \tag{4.8}$$
$$V_{n+1}y_{n+1} = L_nX_n + Dx_D, \tag{4.9}$$

where L, V, and D are the liquid and the two vapor flowrates, respectively.

These balances are the same as for the rectifying section of a continuous column, except that they are time dependent. If constant molal overflow is assumed, L and V become constant, and the operating line is

$$y_{n+1} = \frac{L}{V}x_n + \left(1 - \frac{L}{V}\right)x_D. \tag{4.10}$$

At any time during column operation, this is a line with slope L/V that intersects the diagonal at x_D, x_D. Either the reflux ratio, and thus L/V slope, or x_D will vary during operation, such that the operation line will be constantly changing. If the reflux ratio is varied, McCabe–Thiele analysis can be applied on a stage-by-stage basis to find the relationship between x_D and x_B. The operating line is drawn for a number of x_D values and a specified number of equilibrium stages is stepped off to find the corresponding x_B. Given x_B values for each x_D, Equation (4.5) can be solved by either numerical integration, such as Simpson's rule, or graphical techniques [2].

4.3.2 Operating time

Because it is not a continuous process, the operating time is an important consideration in batch processes. Since the distillate is usually the product, the operating time can be given as a ratio of the total quantity of distillate collected to the distillate flowrate:

$$t = \frac{D_{\text{total}}}{D}, \tag{4.11}$$

where D_{total} is calculated from the Rayleigh equation. The distillate flowrate is based on the maximum vapor, or flooding, velocity for which a column is designed. A mass balance around the condenser gives

$$D_{max} = \frac{V_{max}}{1 + (L/D)}, \tag{4.12}$$

where D_{max} is the distillate flow corresponding to flooding velocity. Columns are usually operated at a fraction of the maximum flowrate [2].

Example 4.1: batch distillation

Problem:

A single-staged batch distillation process is to be used to remove highly volatile organics from water. 1000 kgmol of feed with $x_F = 0.1$ VOCs is charged into a batch still. The water that is removed from the still can be no greater than $x_{B,f} = 0.005$ VOCs. If the relative volatility is 4.3 and does not change appreciably, what quantities of fluid are collected in the bottoms and distillate and what is the VOC composition of the distillate stream?

Solution:

Unknowns are D, B, x_D; solve three equations:

$$F = 1000 = D + B \tag{4.2}$$

$$Fx_F = 0.1(1000) = Dx_D + 0.005B \tag{4.3}$$

$$\ln\left(\frac{B}{1000}\right) = \frac{1}{3.3}\ln\left[\frac{0.005(0.9)}{0.1(0.995)}\right] + \ln\left[\frac{0.9}{0.995}\right]. \tag{4.7}$$

Solving (4.7) $B = 354$ kgmol;
then (4.2) $D = 646$ kgmol;
and (4.3) $x_D = 0.154$.

4.4 Continuous distillation

Most distillation processes are continuous and separate mixtures in which all the feed components are relatively volatile. The feed to a distillation column is usually a liquid. For a vapor feed, the column must be cooled to allow both liquid and vapor flows within the column. This approach would not normally be needed in environmental applications. A distillation column is designed with the feed stream entering somewhere close to the middle where the temperature is above that of the more volatile species and below that of the less volatile ones. A column consists of a number of discrete stages each at a different temperature such that temperature increases from the top of the column to the bottom. The more volatile components in the feed vaporize upon entering the column, rise to the top and are collected as distillate (*D*). The less volatile components remain in the liquid phase, move to the bottom of the column and are usually termed bottoms (*B*). The temperature at the top of the column is just above the boiling point of the most volatile component and that at the bottom is just below that of the least volatile. If the feed stream contains more than two substances to be separated, multiple product streams can be removed down the length of a column. Often, as will be discussed later, a portion of the distillate stream is sent back to the column as reflux.

If the relative volatility of the components to be separated is quite low (close to one), a mass-separating agent (MSA) can be added to alter the phase equilibrium. If the MSA is relatively non-volatile and exists at the bottom of the column, the process is called extractive distillation. Azeotropic distillation occurs when the MSA forms an azeotrope with one or more components in the feed mixture such that the separation is limited to the azeotropic composition.

A fundamental requirement of distillation, as well as all other separations unit operations, is that intimate contact must occur between the phases at each stage in a cascade. In continuous distillation, this means intimate contact between the vapor and liquid phases in each stage. Typical equipment to achieve this requirement is a sieve tray.

A sieve tray, Figure 4.2, consists of a circular horizontal tray, A, with a downpipe, B, which acts as a weir. When the level of liquid sitting on top of each tray becomes too high, it spills over the weir to the next lower tray. The downpipe, C, from the next higher tray reaches nearly to tray A, allowing overflow liquid to travel down to A. It is designed such that the overflow liquid is injected below the surface of the liquid on A. The weirs insure column operation by maintaining a constant liquid level on each sieve tray, regardless of the liquid flowrate through the system.

Across the surface of each sieve tray there are small holes, typically 0.25 to 0.50 inches (6.35 to 13 mm) in diameter. Vapor from the next lower tray flows upward through the

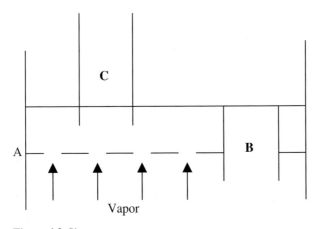

Figure 4.2 Sieve tray.

column via these perforations. Unless the vapor flow is significantly below the operating range, the vapor velocity is adequate to prevent leakage, called weeping, of the liquid through the holes. The perforations also serve the purpose of dividing the vapor into many small bubbles, which pass through the pool of liquid on the trays, thus enhancing the contact of vapor with liquid. The vapor bubbles create a boiling mass of liquid on each tray. Just above this mass and just below the next higher tray is a fog of collapsing bubbles. Some of this fog settles back into the liquid, but some becomes entrained in the vapor and is carried to the plate above. Because of the difference in column operating temperature at each sieve plate, the vapor becomes purer in the more volatile species as it rises, while the liquid becomes more pure in the less volatile species.

4.4.1   McCabe–Thiele analysis

From Chapter 3 we saw that a McCabe–Thiele method of "stepping off" stages could be applied to determine the number of equilibrium stages required to perform a desired separation using only the vapor–liquid equilibrium relationship and a mass balance of the system. Let us begin to apply this technique to distillation by examining the simplest possible scenario. Some simplifying assumptions, therefore, must be made. First, we assume that our feed stream is binary such that only two components, a more volatile and a less volatile, exist throughout the column. Second, the system is operating at steady-state so that all flow rates, compositions, stage temperatures and pressures are constant with respect to time. Third, pressure in the column is assumed to be constant, allowing us to neglect pressure drop through the column. Fourth, it is assumed that there is no chemical reaction between the two components in the system. As a first approximation, the stages are assumed to be ideal; i.e., equilibrium between the liquid and vapor phases is reached at each sieve tray. Methods to account for stage efficiencies will be discussed later in the chapter. Fifth, it is assumed that the column operates adiabatically. This means that heat exchange between the column and the external environment is neglected. This assumption becomes useful in calculating the optimal location of the feed plate.

Figure 4.3 shows a schematic of a binary system distillation column. The x concentrations all refer to the more volatile species in the feed. For analysis the column is divided into two distinct sections: the rectifying section which includes the feed tray and all others above it; and the stripping section which includes all trays below the feed tray. The rectifying section operates to provide an almost pure vapor stream, D, of the more volatile component and the stripping section provides an almost pure liquid stream, B, of the less volatile component. It is important to note that a portion of the vapor stream, or distillate, is condensed and fed back to the top of the column and a portion of the liquid stream, or bottoms, is vaporized and returned to the bottom of the column. The condensed vapor, L, that is returned to the top of the column provides a liquid stream in the rectifying portion of the column. The vaporized liquid from the bottoms that is fed back into the stripping section provides a vapor stream in the stripping section. Without these two recycle streams,

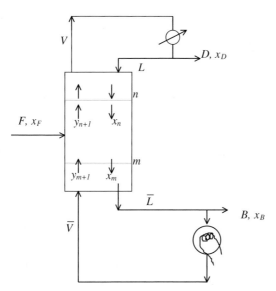

Figure 4.3 Distillation column schematic (see Nomenclature section at end of chapter for variable definition).

a distillation column would operate as a single-stage flash drum with vapor above the feed location and liquid below.

At the top and the bottom of the column there are two options for condensing the vapor and vaporizing the liquid, respectively, prior to recycle. A total condenser at the top of the column simply condenses the entire distillate stream to a liquid. It is not considered to be an equilibrium stage because although there is a complete phase change the distillate composition is not changed. A partial condenser, on the other hand, only condenses a fraction of the distillate such that the compositions of the condensed fraction and the remaining vapor are altered. It is, therefore, considered to be an equilibrium stage in the column. The same description applies to the bottoms stream reboiler. A total reboiler is not an equilibrium stage because the resulting vapor has the same composition as the liquid bottoms stream. A partial reboiler, however, produces both a liquid fraction and a vapor fraction which have different compositions than the entering liquid. Thus, it is an equilibrium stage in the column.

The McCabe–Thiele graphical method uses three important types of lines. First, there is an equilibrium line which gives the vapor–liquid equilibrium relationship of a binary system over a range of concentrations. You can obtain information about azeotropes from this line (where it crosses the diagonal $y = x$ line, vapor and liquid compositions are equal, as shown in Figure 3.7) and the ease of separation which is demonstrated by the distance of the equilibrium line from the diagonal. Second, there are the operating lines which are graphical depictions of the mass balances in each section of the column. There are different operating lines for the rectifying and stripping sections, respectively. These lines relate concentrations of liquid and vapor passing streams between stages: L and V in the rectifying section; and $\overline{L}$ and $\overline{V}$ in the stripping section. Third, there is a q-line which describes the vapor–liquid quality of the feed and is derived with an energy balance. The q-line, to be discussed later in this chapter, occurs at the intersection of two operating lines. The q-line is often called the feed line when it is drawn for a feed stream. The location of the q-line depends on the concentration of the feed and its slope depends on the quality of the feed (the fraction which is liquid). The q-line is used to determine the optimal feed tray in a column.

The McCabe–Thiele analysis begins with deriving the mass balances for the rectifying and stripping sections of the column. Before we begin, an additional assumption of constant molar overflow, CMO, is made. A CMO occurs when the molar flowrates in the liquid and vapor phases remain constant in each section of the column. This assumption corresponds physically to the fact that the molar heats of vaporization for each component in the feed are approximately equal. This means if you condense one mole of vapor, you will create one mole of vapor by evaporation from the liquid phase. This assumption applies to *total* molar flowrates. The one mole of vapor which condenses will normally have a different composition from the one mole of liquid evaporated. In addition, if the total molar flowrates in each of the stripping and rectification sections of the column are constant, a mass balance of the section when plotted as vapor mole fraction vs liquid mole fraction, results in a straight line.

Mass balances

Both component and total mass balances can be performed using the entire column as the control volume. The total and component for the more volatile species are:

Overall: $\qquad\qquad\qquad F = D + B$ $\qquad\qquad\qquad\qquad\qquad$ (4.2)

Component A: $\qquad x_F F = x_D D + x_B B,$ $\qquad\qquad\qquad\qquad$ (4.3)

where x refers to the more volatile component and the subscripts F, D and B refer to its respective concentrations in the feed, distillate and bottoms streams.

Upper (rectifying) section – enrichment of more volatile component

First, remember the assumption that liquid and vapor flowrates in the rectifying section of the column are constant. An overall mass balance can then be written around the top of the column:

$$V = D + L \quad \Rightarrow \quad D = V - L. \qquad\qquad\qquad\qquad (4.13)$$

In addition, a component mass balance can be performed which describes the net flowrate of component A in the upper section for any tray. For clarification, trays in distillation columns are typically numbered from the top to the bottom, such that the top tray is stage 1 and the bottom is stage m. Thus, liquid flows down the column from tray n to tray $n + 1$

and vapor flows up from tray $n + 1$ to tray n. The component mass balance, then, becomes

$$x_D D = y_{n+1} V - L x_n \qquad (4.14)$$

for any two subsequent trays.

Because the goal of the mass balance is to obtain an operating line describing the relationship between the vapor-phase mole fraction entering and the liquid-phase mole fraction leaving a stage, the mass balances can be combined and rearranged to give:

$$y_{n+1} = \frac{L}{V} x_n + \frac{x_D (V - L)}{V} \qquad (4.15)$$

or, in terms of liquid streams only,

$$y_{n+1} = \frac{L}{L + D} x_n + \frac{x_D D}{L + D}. \qquad (4.16)$$

Reflux ratio

Typically, the quantity of distillate product that is being condensed and returned to the top of the column needs to be specified. It can be defined as either an internal or external quantity depending upon which variables are used. The internal reflux ratio is defined in terms of flowrates within the rectifying section of the column:

$$R_V = \frac{L}{V} = \text{reflux ratio (internal).} \qquad (4.17)$$

The external reflux ratio is expressed in terms of liquid flows exiting the condenser:

$$R_D = \frac{L}{D} = \text{reflux ratio (external).} \qquad (4.18)$$

The operating line for the rectifying section can, then, be expressed in terms of the external reflux ratio. It is chosen because of the ease of measuring and controlling external liquid flowrates compared to internal vapor ones:

$$y_{n+1} = \frac{R_D}{1 + R_D} x_n + \frac{x_D}{1 + R_D} \qquad \text{(operating line for the rectifying section).} \qquad (4.19)$$

So, with a single point on the x–y diagram and a reflux ratio, the operating line can be drawn. Let us look at what happens when the liquid coming from a total condenser equals the distillate concentration. Substituting x_D for x_n in the operating-line equation,

$$y = \frac{L}{L + D} x_D + \frac{D x_D}{L + D} = x_D, \qquad (4.20)$$

it can be seen that at the top of the column $y = x_D$ or the operating line intersects the diagonal at (x_D, x_D). So, if you know x_D and R_D, you can draw the operating line for the rectifying section. It is important to note that since there is more vapor than liquid in the rectifying section (some vapor is removed as D), *the slope of this operating line is LESS than 1. Mathematically this is apparent because the slope of $R_D/(R_D + 1)$ is less than 1.*

Example 4.2: enriching column

Problem:

Steam stripping is to be used to remove a solvent from contaminated soil. An enriching column will be used to recover the solvent from the stream. A vapor feed of 40 mol/hr with a composition of 20 mol% solvent and 80 mol% water enters an enriching column. The distillate stream is to have a flow rate of 5 mol/hr and a concentration of 90 mol% solvent. The internal reflux ratio is 0.875 and constant molar overflow (CMO) may be assumed. Graph the operating line to predict the number of equilibrium stages in this enriching column.

Solution:

Assuming CMO means that L and V are constant, and the operating line will be straight. The internal reflux ratio is L/V, so the easiest way to plot the mass balance for the column (operating line) is to use the Equation (4.10):

$$y_{n+1} = \frac{L}{V}x_n + \frac{x_D(V - L)}{V}.$$

The slope of the operating line is the internal reflux ratio, 0.875. Now that we know the slope, we need only one point to plot the line. It is possible to plot the y-intercept of this line, but an easier point to find is the one where the operating line crosses the diagonal, at $x_D = 0.9$ (liquid and vapor compositions are equal since a *total* condenser is used, i.e., <u>all</u> vapor is condensed to liquid).

Now we can solve the two mass balances simultaneously to find the composition of the stream leaving the bottom of the column:

$$F = D + B \Rightarrow 40 \,\frac{\text{mol}}{\text{hr}} = 5 \,\frac{\text{mol}}{\text{hr}} + B$$

$$x_F F = x_D D + x_B B \Rightarrow 0.20 \,\frac{\text{moles solvent}}{\text{moles feed}}\left(40 \,\frac{\text{mol}}{\text{hr}}\right)$$

$$= 0.90 \,\frac{\text{moles solvent}}{\text{moles distillate}}\left(5 \,\frac{\text{mol}}{\text{hr}}\right) + x_B B.$$

So the bottoms have a concentration of 0.10 mol% solvent and a flowrate of 35 mol/hr. Once this concentration is found on the operating line, the stages can be "stepped off" (Figure 4.4). Starting at x_D, draw a horizontal line to the equilibrium curve. This point will give the mole fractions of the vapor and liquid streams on the first stage. From this point, draw a vertical line down to the operating line. This point will give the mole fractions of the passing streams (x_n and y_{n+1}, $n = 1$ for the passing streams below the first stage). The first "step" is now complete, and the process is repeated until the passing streams from the last stage meet or exceed the required value for x_B. Remember that the points on the operating line are passing stream mole fractions, while the points on the equilibrium line are the mole fractions of the streams exiting each equilibrium stage. Now the number of stages can be counted. Whenever the desired concentration falls between two stages, the larger value is used. In this example, three stages will not quite achieve the desired separation, but four stages will reduce the bottoms concentration to less than 0.10 mol% solvent, which does not work with the mass balances solved earlier. It is important to note that the graphical method is not exact, but it is much easier than stage-by-stage calculations. The answer to the problem by the McCabe–Thiele method is four equilibrium stages, but note that the concentrations produced by such a column will probably not be equal to those specified in the problem statement (the column should exceed specifications if it is operating efficiently).

The question then comes up of what do we adjust to change the design. The value of x_D is fixed so a change in the operating line means a change in slope. How would we accomplish this? What is the limit? Why?

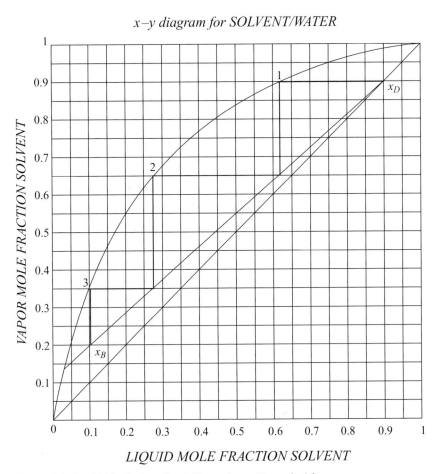

x–y diagram for SOLVENT/WATER

VAPOR MOLE FRACTION SOLVENT

LIQUID MOLE FRACTION SOLVENT

Figure 4.4 Graphical solution of enriching column, Example 4.2.

Lower (stripping) section

The portion of the column that lies below the feed plate is called the stripping section. The function of this section is to obtain a nearly pure bottoms stream, B, of the less volatile component of the feed. As with the rectifying section, we can analyze in terms of a mass balance to obtain an operating line relating the vapor-phase mole fraction to the liquid-phase mole fraction exiting a stage. Again, it is assumed that the molar flowrates B, L and V are constant and that constant molar overflow applies. The overall mass balance around the bottom of the column, then, becomes:

$$\overline{L} = B + \overline{V} \quad B = \overline{L} - \overline{V}, \tag{4.21}$$

and the component mass balance of the more volatile species around any stage m is:

$$x_B B = x_m \overline{L} - (\overline{L} - B) y_{m+1}. \tag{4.22}$$

Combining these two equations and solving for the vapor mole fraction in terms of process variables gives:

$$y_{m+1} = \frac{\overline{L}}{\overline{L} - B} x_m - \frac{B x_B}{\overline{L} - B}, \tag{4.23}$$

or

$$y_{m+1} = \frac{\overline{L}}{\overline{V}} x_m - \frac{x_B(\overline{L} - \overline{V})}{\overline{V}}. \tag{4.24}$$

Similarly, with a single point and the $\overline{L}/\overline{V}$ ratio, a line describing mass balance in the stripping-section of the column can be drawn. To obtain the point, let us look at what happens when the liquid exiting the bottom of the column is at its bottoms concentration, x_B. Substituting x_B into the stripping-section mass balance, we see that $y = x_B$ or the stripping line intersects the diagonal at (x_B, x_B) at the bottom of the column. Hence, if L and V are known in addition to the bottoms concentration of the more volatile species, a stripping line can be drawn. As before, it is important to note that since there is more liquid than vapor in the stripping section (liquid removed in bottoms), *the slope of this operating line is GREATER than 1.*

Stripping columns also exist, just like the enriching, or rectifying, column in the previous example. The purpose of a stripping column would be to purify the least volatile component; an enriching column would purify the most volatile component.

Macroscopic energy balance

In distillation, the separating agent is energy. Any complete analysis of a distillation system would include energy balances. For a distillation column with both stripping and enriching sections, the energy balance must take into account the enthalpies of all entering and exiting streams and the energy requirements, or duties, of the condenser and the reboiler.

Remember that this balance assumes that the system operates adiabatically such that no heat gains or losses occur between the distillation column and its surroundings.

Referring to Figure 4.3, we can evaluate the thermal energy inputs and outputs. At steady-state, just as with mass balances, the energy inputs (gains) must equal the outputs (losses):

$$Fh_F + Q_R = Dh_D + Bh_B + Q_C, \tag{4.25}$$

where F, D, and $B =$ molar flowrates of the feed, distillate, and bottoms
$\quad h_F$, h_D, and $h_B =$ molar enthalpies of these streams
$\quad\quad Q_C =$ energy requirement of the condenser
$\quad\quad Q_R =$ energy requirement of the reboiler.

The enthalpies can be determined from an enthalpy–composition diagram or from the heat capacities and latent heats of vaporization. In the case of a total condenser, the composition of the stream is not changed (not true in a partial condenser!) and a mass balance around the condenser can be written:

$$V = L + D, \tag{4.26}$$

or, in terms of the external reflux ratio (R_D), as

$$V = \frac{L}{D}D + D = (1 + R_D)D, \tag{4.27}$$

and, with a given value of R_D, V can be calculated.

The condenser energy balance is:

$$VH = (D + L)h_D + Q_C, \tag{4.28}$$

where D and L are at the same conditions, and H is the enthalpy of the vapor stream going to the condenser or, combining equations,

$$Q_C = V(H - h_D). \tag{4.29}$$

Substitution then gives:

$$Q_C = (1 + R_D)D(H - h_D). \tag{4.30}$$

Now this value can be used with the overall energy balance for the entire column to find the reboiler duty, Q_R.

Feed line

As we have seen, the McCabe–Thiele diagram for a distillation column with rectifying and stripping sections will have two operating lines, usually referred to as top and bottom operating lines, respectively. When stepping off stages from the top, you use the top operating line (mass balance) until the mass balance changes. This will occur when feed enters the column. You want the feed to be introduced such that its composition and

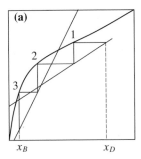 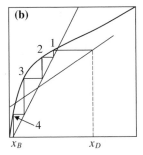

Figure 4.5 Feed stage location is (a) optimum, and (b) above optimum.

thermal condition are a good match to the column conditions on that stage (see Figure 4.5). This point is the optimal feed stage; a separation will require the fewest total number of stages when the optimum feed stage is used as the actual feed stage. If the feed is not at the optimum stage location, the point at which you change operating lines is different. The actual feed plate location is where the "steps" cross over from one operating line to the other since the mass balance changes at that location. Under optimum conditions, one always uses the operating line that is furthest from the equilibrium line.

The phase and temperature of the feed will affect the vapor and liquid flowrates in the column. The mass and energy balances around the feed plate are:

$$F + \overline{V} + L = \overline{L} + V$$

$$F h_F + \overline{V} H_{f+1} + L h_{f-1} = \overline{L} h_f + V H_f, \tag{4.31}$$

where h_F is the enthalpy of the feed, and f is the location of the feed plate. Assuming CMO means that the liquid and vapor enthalpies don't vary much throughout the column. Solving the mass balance for $\overline{V} - V$ and substituting into the energy balance gives:

$$\frac{\overline{L} - L}{F} = \frac{H - h_F}{H - h} = q = quality, \tag{4.32}$$

or, in words,
(vapor enthalpy on feed plate – feed enthalpy)/(vapor enthalpy on feed plate – liquid enthalpy on feed plate) = quality = the fraction of feed that is liquid.

Now remember that the feed plate is where the operating lines intersect each other. Solving mass balances for the top and bottom sections of the column gives:

$$\begin{aligned} yV &= Lx + Dx_D \\ y\overline{V} &= \overline{L}x - Bx_B \end{aligned} \qquad \text{gives:} \qquad y = -\frac{\overline{L} - L}{\overline{V} - V}x + \frac{Fx_F}{\overline{V} - V}. \tag{4.33}$$

Since each term except y and x is constant, this is a straight line called the feed line.

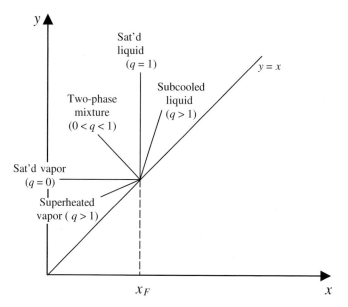

Figure 4.6 Feed lines.

Solving the mass balance around the feed plate for $V - \overline{V}$ and substituting into this feed-line equation gives:

$$y = -\frac{\overline{L} - L}{F - (\overline{L} - L)}x + \frac{Fz_F}{F - (\overline{L} - L)} = -\frac{(\overline{L} - L)/F}{F/F - \left(\dfrac{\overline{L} - L}{F}\right)}x$$

$$+ \frac{F/Fz_F}{F/F - \left(\dfrac{\overline{L} - L}{F}\right)} = \frac{q}{q - 1}x + \frac{1}{1 - q}z_F. \tag{4.34}$$

So, the slope of the feed line is found as $q/(q - 1)$. If the feed is a saturated liquid, then $q = 1.0$, the slope of the line is infinity and the feed line is vertical. Conversely, if the feed is a saturated vapor, the slope is zero and the feed line is horizontal. The feed can also exist as a two-phase mixture, subcooled liquid, and superheated vapor. Figure 4.6 presents a number of representative feed lines and Figure 4.7 shows different qualities of feed entering a distillation column.

The result above shows that the operating lines above and below the feed line, and the feed line itself, all intersect at a common point.

As pointed out at the beginning of this chapter, the coolest temperature is at the top of the column and increases as you proceed down to the reboiler. The slope of the feed line reflects this as the cooler feeds point toward the top (x_D) and the hottest feeds point toward the bottom (x_B).

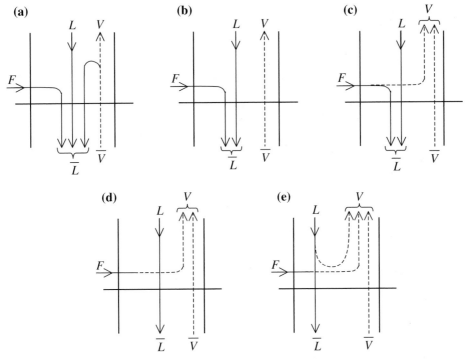

Figure 4.7 Different types of feeds. q = moles of liquid flow in stripping section that result from one mole of feed. (a) $q > 1$: cold liquid feed; (b) $q = 1$: feed at bubble point (saturated liquid); (c) $0 < q < 1$: feed partially vapor; (d) $q = 0$: feed at dew point (saturated vapor); (e) $q < 0$: feed in superheated vapor.

To plot the feed line, you will need a point in addition to the slope. The y-intercept could be this point, but an easier one to find is where the feed line intersects the diagonal at $y = x = x_F$.

The analysis for the feed stream can be generalized and applied to any side stream, whether it is a feed or a withdrawal. Withdrawal streams will be discussed later (see *Side streams*, p. 111).

Example 4.3: complete McCabe–Thiele method

Problem:
We desire to use a distillation column to separate an ethanol–water mixture. The column has a total condenser, a partial reboiler, and a saturated liquid reflux. The feed is a saturated liquid of composition 0.10 mole fraction ethanol and a flow rate of 250 mol/hr. A bottoms mole fraction of 0.005 and a distillate mole fraction of 0.75 ethanol is desired. The external reflux ratio is 2.0. Assuming constant molar overflow, find the flowrates, the number of equilibrium stages, optimum feed plate location, and the liquid and vapor compositions leaving the fourth stage from the top of the column. Pressure is 1 atm.

Solution: (see Figure 4.8)

Given:

$F = 250$ mol/hr

$x_F = 0.10$

$x_B = 0.005$

$x_D = 0.75$

$R_D = 2.0$.

Mass balances around the entire column:

$$F = D + B \Rightarrow 250 \; \frac{\text{mol}}{\text{hr}} = D + B$$

$$x_F F = x_D D + x_B B \Rightarrow 0.10 \; \frac{\text{moles EtOH}}{\text{moles feed}} \left(250 \; \frac{\text{mol}}{\text{hr}} \right)$$

$$= \left(0.75 \; \frac{\text{moles EtOH}}{\text{moles distillate}} \right) D + \left(0.005 \; \frac{\text{moles EtOH}}{\text{moles bottoms}} \right) B.$$

Two equations and two unknowns gives $B = 218$ mol/hr and $D = 32$ mol/hr.

Top operating line:

$$\text{Slope} = \frac{L}{V} = \frac{L/D}{L/D + 1} = \frac{R_D}{R_D + 1} = \frac{2}{2+1} = \frac{2}{3};$$

one point $= y = x = x_D = 0.75$.

Feed line:

$$\text{Slope} = \frac{q}{q - 1} = \frac{1}{1 - 1} = \infty = \text{vertical line};$$

one point $= y = x = x_F = 0.10$.

Bottom operating line:

First point $=$ intersection of feed and top operating line;

Second point $= y = x = x_F = 0.005$.

Stepping off stages:

The separation will take eight equilibrium stages (the last stage will be the reboiler) and the optimum feed plate will be the fifth from the top. The liquid composition on the fourth stage from the top is ~0.28 mole fraction ethanol, and the vapor composition on this stage is ~0.56 mole fraction ethanol.

Minimum reflux

Minimum reflux is defined as the external reflux ratio, R_D, where the desired separation could be obtained with an infinite number of stages. Obviously, this is not a real condition, but the concept is useful because actual reflux ratios are often defined in terms of minimum reflux ratios.

The number of stages becomes infinite at a "pinch point" on a McCabe–Thiele diagram, as shown in Figure 4.9. A "pinch point" occurs at the intersection of the operating and

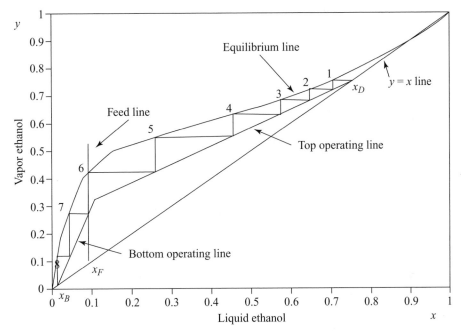

Figure 4.8 Complete McCabe–Thiele analysis of a distillation process.

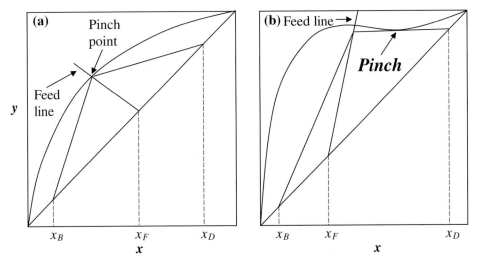

Figure 4.9 Minimum reflux analysis for pinch points (a) at, and (b) above the feed plate.

equilibrium lines. Since there is no gap between the lines, it is not possible to step off stages beyond this point. This usually occurs at the intersection of the operating, q, and equilibrium lines. Some systems, however, do not have pinch points at this intersection due to the shape of the vapor–liquid equilibrium line.

The minimum reflux ratio (L/D_{min}) corresponds to the lowest possible value of the slope of the top operating line. On a McCabe–Thiele diagram this operating line begins at the point $y = x = x_D$, and ends at the point where the feed and equilibrium lines intersect. The slope of this operating line can then be used to find the minimum reflux ratio.

In some cases, the operating line may intersect the equilibrium line below the intersection of the q and equilibrium lines. In general, the operating line corresponding to minimum reflux can be constructed by rotating the line upward about the point $y = x_D$ until the line contacts the equilibrium line at or above the feed line. [Remember: slope of top operating line $= R_D/(R_D + 1)$.]

Total reflux

Total reflux is similar to minimum reflux in that it is not usually a real condition. In total reflux, all of the overhead vapor is returned to the column as reflux, and all of the liquid is returned as boilup, so that there are no distillate and bottom flows out of the column. At steady-state, this means that the feed stream flowrate is also zero. Total reflux is used in actual columns during start up and also to test their efficiency. Total reflux is useful in a McCabe–Thiele analysis in order to find the minimum number of stages required for a given separation.

Since all of the vapor is refluxed, $L = V$ and therefore $L/V = 1$. The same is true in the stripping section, so that $\overline{L}/\overline{V} = 1$. So both operating lines are now the $y = x$ line. This is shown in Figure 4.10. The minimum number of stages required is determined by stepping off stages using the equilibrium line and the diagonal $(y = x)$ line.

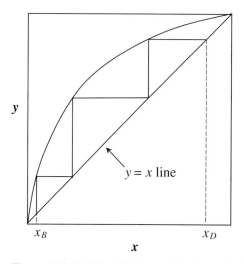

Figure 4.10 McCabe–Thiele analysis for total reflux and minimum number of plates.

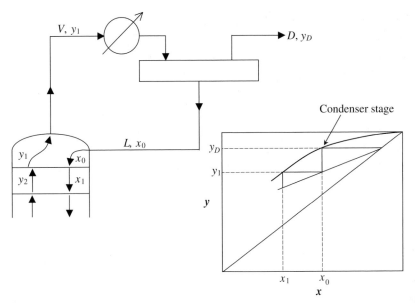

Figure 4.11 Partial condenser.

Partial condensers

A partial condenser, as shown in Figure 4.11, condenses only a portion of the overhead stream and returns it as reflux, so the distillate product remains vapor. There are only two things which make the analysis of a column with a partial condenser different from one with a total condenser. The first is the point where the top operating line intersects the diagonal on a McCabe–Thiele diagram. The point is still the same, but the notation x_D is no longer useful because the distillate is vapor. Instead, y_D is used and the operating line is plotted as before. The other thing to note is that while stepping off stages, *a partial condenser is counted as one equilibrium stage, while a total condenser is not.* A partial condenser has both a vapor and liquid exit stream. They are assumed to be in equilibrium, so the compositions are different from the single vapor stream entering the condenser. A total condenser, on the other hand, has a vapor at a given composition entering and only liquid at the same composition exits (i.e., no vapor–liquid equilibrium in the exit stream).

Example 4.4: column with partial condenser (Figure 4.12)

Problem:
A 100 mol/hr feed stream containing 30 mol% of a contaminant A is to be distilled in a column consisting of a reboiler, one plate, and a partial condenser. Each can be considered as an ideal stage. To concentrate the contaminant prior to further treatment, the distillate should contain 80 mol% A. The ratio of liquid reflux flowrate to distillate flowrate is 2.

Calculate: L, D, and the composition of A in the L stream coming off the reboiler. Use a McCabe–Thiele diagram. Be sure to label the equilibrium curve and the operating line.

Solution (see Figures 4.13 and 4.14):

Definition of a partial condenser: y_D and x_1 are in equilibrium:

The material balances give a slightly different answer than that obtained from the McCabe–Thiele analysis shown in Figure 4.15. The McCabe–Thiele analysis is only an approximate solution, and it may be that the discrepancy comes from the need for more equilibrium data points to obtain a more accurate equilibrium curve.

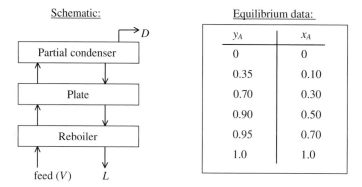

Figure 4.12 Schematic and equilibrium data for partial condenser, Example 4.4.

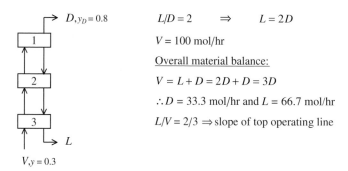

Figure 4.13 Equilibrium stages and passing streams, Example 4.4.

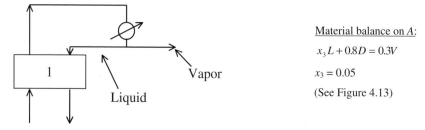

Figure 4.14 Partial condenser on Stage 1, Example 4.4.

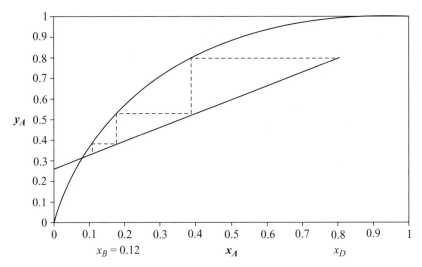

Figure 4.15 Stepping off stages for partial condenser, Example 4.4.

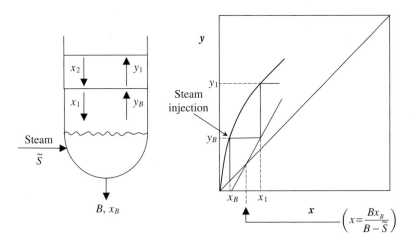

Figure 4.16 Steam injection.

Steam injection

Steam distillation, as shown in Figure 4.16, uses steam injection instead of a reboiler to provide heat. One advantage is a lower operating temperature. This approach is often preferable for wastewater treatment (steam stripping) when the contaminant concentration in the exit water stream must be very low. The operating line for the stripping section is still valid:

$$y_m = \frac{\overline{L}}{\overline{V}}x_{m+1} - \frac{Bx_B}{\overline{V}}.$$

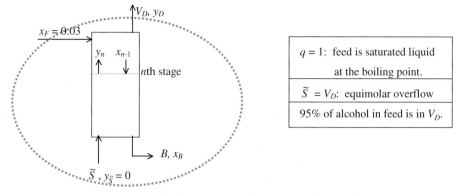

Figure 4.17 Stripping tower with direct steam injection (dotted line indicates control volume), Example 4.5.

The overall mass balance on the lower section is:

$$\overline{L} + \widetilde{S} = \overline{V} + B \Rightarrow \overline{L} - \overline{V} = B - \widetilde{S}. \tag{4.35}$$

The intersection of the operating line with the diagonal ($y = x$ line):

$$x = \frac{Bx_B}{B - \widetilde{S}} \quad \text{(not } x_B\text{)}.$$

The intersection with the x-axis ($y = 0$): $\qquad x = \dfrac{Bx_B}{\overline{L}}.$

But when assuming constant molar overflow:

$$\widetilde{S} = \overline{V} \quad \therefore \quad B = \overline{L} \quad \therefore \quad x = x_B \quad \text{when} \quad y = 0.$$

Stepping off stages: use $(x_B, 0)$ as an end point of operating line.

Example 4.5: stripping tower and direct steam injection

Problem:

A liquid ethanol–water feed at the boiling point contains 3 mol% ethanol and enters the top tray of a stripping tower, as shown in Figure 4.17. Saturated steam is injected directly into the liquid in the bottom of the tower. 95% of the alcohol in the feed exits in the overhead. Assume equimolar overflow for this problem and an equilibrium relationship:

$$y = 8.3x.$$

(a) Calculate the minimum moles of steam needed per mole of feed.
(b) Using 2.5 times the minimum moles of steam, calculate the number of theoretical stages needed and the composition of the overhead and the bottoms streams.

Solution:

(a) Mass balance (basis $F = 100$ mol/hr) on entire column:

$$F + \tilde{S} = V_D + B \quad \text{(overall)}$$

$$F = B \quad (\text{since } \tilde{S} = V_D)$$

$$x_F F + y_S \tilde{S} = y_D V_D + x_B B$$

$$y_D V_D = 0.95 x_F F$$

$$x_B = \frac{0.05 x_F F}{B} = \frac{0.05 x_F F}{F} = 0.05 x_F = 1.5 \times 10^{-3}.$$

Mass balance around control volume:

$$(0)\tilde{S} + x_{n-1} B = x_B B + y_n \tilde{S}$$

$$y_n = \frac{B}{\tilde{S}} x_{n-1} - \frac{B}{\tilde{S}} x_B.$$

$\therefore$ *Minimum $\tilde{S}$ corresponds to the maximum slope of the operating line.*

To determine the maximum value of $B/\tilde{S}$, substitute values for y_n and x_{n-1} which correspond to intersection of operating and equilibrium line. Rearranging the mass balance equation:

$$8.3 \, (0.03) = \frac{B}{\tilde{S}}(0.03) - \frac{B}{\tilde{S}}(1.5 \times 10^{-3})$$

$$\frac{\tilde{S}}{B} = \frac{\tilde{S}}{F} = \frac{0.03 - 1.5 \times 10^{-3}}{0.27} = 0.11 \text{ moles steam}_{\min}/\text{moles feed.}$$

(b) Actual steam rate $= 2.5$ times the minimum.

$$\therefore \tilde{S} = 2.5(0.11)\left(100 \frac{\text{mol}}{\text{hr}} \text{feed}\right) = 27.5 \frac{\text{mol}}{\text{hr}}$$

$$\therefore \left(\frac{B}{\tilde{S}}\right)_{\text{actual}} = \frac{100 \text{ mol/hr}}{27.5 \text{ mol/hr}} = 3.6.$$

Actual operating line is:

$$y = 3.6x - 3.6(1.5 \times 10^{-3}) = 3.6x - 5.5 \times 10^{-3}.$$

The line can be graphed by plotting two points. We already know one set of points: $(0, 1.5 \times 10^{-3})$. The other will be at $x = 0.03$:

$$y = 3.6(0.03) - 5.5 \times 10^{-3} = 0.10.$$

From Figure 4.18, three theoretical stages are required.

Side streams (see Figure 4.19)

1 The operating line for the top and bottom sections are still drawn as before.
2 The q-line for the feed is still drawn as before.
3 To locate the middle operating line x_S must be known. The middle operating line begins at the top operating line at the point $x = x_S$ and ends at the point where the q-line for the feed intersects the bottom operating line.

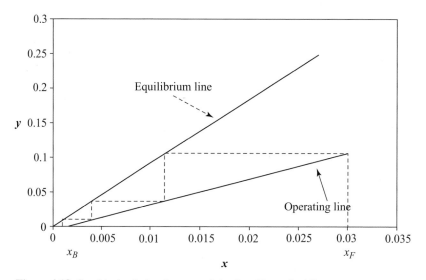

Figure 4.18 Graphical solution for steam injection, Example 4.5.

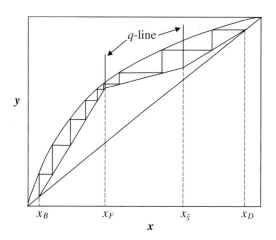

Figure 4.19 McCabe–Thiele analysis for a system with a side stream.

Note 1: If an equilibrium stage does not fall exactly on x_S you cannot physically obtain that side-stream composition. You must adjust the reflux ratio until this is corrected.

Note 2: You can treat multiple feeds in a similar manner.

Withdrawal streams are negative feed streams:

$$q = \frac{\overline{L} - L}{F} = \frac{L' - L}{-S} = \frac{L - L'}{S}.$$

If S is a saturated liquid, $q = 1$. If S is a saturated vapor, $q = 0$.

$\therefore q/(q - 1)$ can be used as the slope of the withdrawal line.

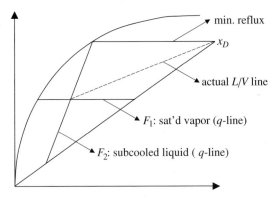

Figure 4.20 When *q*-lines cross.

What if the feed lines cross?

The *minimum reflux* ratio is still determined by the intersection of the operating line from x_D with the *q*-line which intersects the equilibrium line furthest up the column.

Stepping off stages: starting from x_D, step off stages until you reach a *q*-line, then shift to new operating line for optimum feed location. For Example 4.5 (Figure 4.20), F_1 would be introduced above F_2 in the column.

Stage efficiencies

In the McCabe–Thiele analysis thus far, ideal equilibrium stages, or those with perfect efficiencies, have been assumed. An ideal stage is one in which thermodynamic equilibrium between vapor and liquid phases entering each stage is reached before the streams exit the stage. Non-ideal stages, or those with less than perfect efficiencies, do not reach thermodynamic equilibrium. Two approaches exist for determining the number of real stages required to perform a separation as a function of the number of ideal stages.

The first method uses an overall efficiency, η:

$$\eta = \frac{\text{theoretical (equilibrium) stages}}{\text{actual stages}} \times 100(\%).$$

The number of real stages, then, is merely the number of ideal stages divided by the overall efficiency. While this method is simple to use, the difficulty is in obtaining a value for the overall efficiency of a column. In addition, the efficiency often varies from stage to stage, making a single overall value somewhat meaningless. Often, if efficiencies of individual stages can be estimated, they are averaged to give an overall value.

The second method used is the Murphree vapor efficiency:

$$E_{\text{MV}} = \frac{y_n - y_{n-1}}{y_n^* - y_{n-1}} \times 100(\%),$$

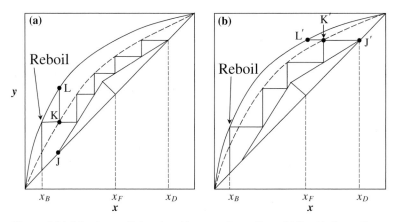

Figure 4.21 Murphree efficiencies: (a) vapor phase, E_{MV}, (b) liquid phase, E_{ML}.

where y_n = actual composition of V phase leaving the stage

   y_n^* = composition of hypothetical V phase that would be in equilibrium with L phase leaving actual stage.

This method is more meaningful than the overall efficiency because it is based on the difference between true operating and hypothetical equilibrium vapor-phase concentrations. However, Murphree efficiencies also vary from stage to stage and require difficult measurement of process variables.

   The Murphree efficiency is the ratio of distance between the operating line and the equilibrium line ($= [\overline{JK}/\overline{JL} \times 100]\%$ in Figure 4.21(a)). Similarly, the Murphree liquid efficiency is given by:

$$E_{ML} = \frac{x_n - x_{n+1}}{x_n^* - x_{n+1}} \times 100(\%),$$

   ($= [J'K'/J'L' \times 100]\%$ in Figure 4.21(b)).

Note: Remember that a reboiler is an equilibrium stage, even though the other stages in a column will not reach equilibrium if they are not 100% efficient. Therefore, the last stage (the one which gives x_B) will appear on the graph at the solid equilibrium curve, not on the dashed curve. The dashed curve is a "pseudo-equilibrium" which describes the two phases when they don't have a chance to completely reach equilibrium.

Example 4.6: plutonium stabilization at Los Alamos National Laboratories [4]

Problem:

Currently, the nitric acid used in plutonium stabilization operations at a particular facility is evaporated to remove dissolved solids, assayed for radioactive content and then sent via underground pipe to a low-level waste handling facility. The acid stream is then neutralized with caustic to remove radioactivity and the resulting solids are immobilized in cement as a TRU (trans-uranic waste). The filtrate is then sent for further processing to meet the current permit standards. The liquid stream, once treated to remove radioactivity, is then discarded.

A new permit with the state of New Mexico will require this liquid stream to meet the domestic water supply limit for dissolved NO_3 of 10 mg/L measured as N (in order to comply with the federal *Clean Water Act* and the *New Mexico Water Quality Act*). A nitric-acid fractionation (distillation column) is to be installed to recover these nitrates from the discard stream prior to disposal. The column (Figure 4.22) will include a total condenser and partial reboiler and will operate with a reflux ratio of 0.4.

A 1.0 L/min stream of a 2 M (22.3 wt% acid) HNO_3 solution will be fed to a preheater prior to entering the distillation column. The preheater will raise the temperature of the feed to just below its bubble point. The stream which is to be disposed will be treated to less than 0.007 M HNO_3 (450 ppm). The bottoms from the column must be concentrated to 12 M (62.9 wt% acid) HNO_3 before it can be reused in the plutonium stabilization process. An azeotrope exists at 15.6 M HNO_3 (45 wt%). Find the number of equilibrium stages in the column, assuming the overall efficiency is about 0.7.

Solution:

The equilibrium data are given in Figure 4.23. Note that the mole fractions are H_2O, not HNO_3. The concentrations given above are converted to mole fractions of H_2O using the mass flowrates listed in the summary sheet from LANL (Figure 4.24). For example, the mole fraction of water in the feed, x_F, is found:

$$\frac{52.932 \text{ kg } H_2O}{hr} \frac{\text{kgmol}}{18.01 \text{ kg}} = 2.94 \text{ kgmol } H_2O/hr$$

$$\frac{15.155 \text{ kg } HNO_3}{hr} \frac{\text{kgmol}}{63.01 \text{ kg}} = 0.24 \text{ kgmol } HNO_3/hr.$$

$$\therefore x_F = \frac{2.94}{2.94 + 0.24} = 0.92.$$

Similarly, $x_D = 1$, and $x_B = 0.67$.

Feed line

One point on the feed line is $y = x = x_D$. Since the feed is a saturated liquid near its bubble point, $q = 1$ and the slope of the feed line is infinity.

Top operating line

One point on the top operating line is $y = x = x_D$. The slope is

$$\frac{L/D}{L/D + 1} = \frac{R_D}{R_D + 1} = \frac{0.4}{0.4 + 1} = 0.29.$$

Bottom operating line

A point on the bottom operating line is $y = x = x_B = 0.67$. The line also goes through the point where the top operating line and the feed line cross. Alternatively, the slope of the bottom operating line could have been found as:

$$\frac{L + F}{V} = \frac{L + F}{L + D} = \frac{L/D + F/D}{L/D + 1} = \frac{0.4 + 3.179/2.442}{0.4 + 1} = 1.21.$$

Figure 4.23 reveals that there are four-and-a-half steps on the graph, so five equilibrium stages will be used. With an efficiency of 0.7, seven equilibrium stages will be required. Note that since x_D is so close to the point where the $y = x$ and equilibrium lines meet, the accuracy in the graphing technique here is low, and anywhere from five to eight steps could be considered a correct answer.

The 12 M HNO_3 stream which will be re-used for plutonium stabilization will reduce the cost of the process; currently the nitrates are discarded as waste. The savings in waste storage and disposal costs is estimated to be US $100,000 annually.

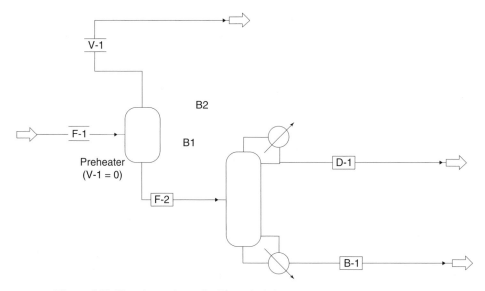

Figure 4.22 Flowsheet schematic, Example 4.6.

4.5 Remember

Distillation uses an energy-separating agent: heat. It is an equilibrium-limited separation. Vapor–liquid equilibrium (VLE) is the type of equilibrium by which distillation columns separate, and vapor pressure is the primary property difference that forms the basis for separation.

- Distillation is a UNIT OPERATION. Regardless of what chemicals are being separated, the basic design principles for distillation are always similar.
- The assumption that stages in a distillation column are in equilibrium allows calculations of concentrations and temperatures without detailed knowledge of flow patterns and heat and mass transfer rates. This assumption is a major simplification.

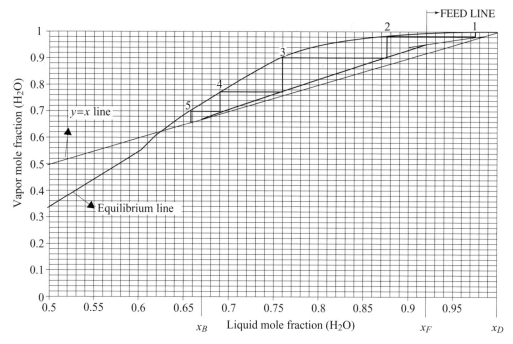

Figure 4.23 LANL, Example 4.6: HNO_3 and H_2O.

- A reboiler and partial condenser are each an equilibrium stage.
- In general, if no azeotropes or side streams are involved, you can separate n products with $n - 1$ columns.
- It is necessary that vapor and liquid compositions are different at the equilibrium conditions that one plans to use (i.e., no azeotrope), otherwise no separation will occur beyond the azeotropic condition.
- Distillation is different from evaporation because both components in distillation are appreciably volatile. In evaporation, usually only one component is vaporized.
- The products need to be thermally stable over the temperature range of operation.
- No components which react exothermally (i.e., generate heat) should be present. These reactions can "run away" and form explosive conditions.
- The McCabe–Thiele graphical method uses three important types of line: the operating lines, an equilibrium line and a q-line (feed line).

4.6 Questions

4.1 Show that higher relative volatilites correspond to a greater distance between the vapor–liquid equilibrium line and the $y = x$ line.

4.2 Show that the result of Question 4.1 corresponds to easier separation by distillation.

===

Data file created by ASPEN PLUS Rel. 9.1–3 on 14 : 18 : 16 Thu Sep 14, 1995
Run ID: HNO3SN-_ Item: STREAM-SUM Screen: Stream-Sum.Main
C----------------C----------------C----------------C----------------C----------------C----------------C----------------

Display ALLSTREAMS		B-1	D-1	F-1	F-2	V-1
Units:		From B1	B1		B2	B2
Format: ELEC-M		To		B2	B1	
	Phase	LIQUID	LIQUID	LIQUID	LIQUID	MIXED
Temperature	[K]	388.7	366.6	298.1	371.6	
Pressure	[ATM]	0.9	0.8	0.8	0.8	0.8
Vapor Frac		0.000	0.000	0.000	0.000	
Solid Frac		0.000	0.000	0.000	0.000	
Mole Flow	[KMOL/HR]	0.737	2.442	3.179	3.179	0.000
Mass Flow	[KG/HR]	24.089	43.998	68.087	68.087	0.000
Volume Flow	[L/MIN]	0.320	0.762	1.000	1.041	0.000
Enthalpy	[MMKCAL/HR]	−0.043	−0.164	−0.212	−0.208	
Mass Flow	[KG/HR]					
H2O		8.396	43.996	52.932	52.932	
HNO3		25.153	0.002	15.155	15.255	
H3O+						
NO3−						
OH−						

Display ALLSTREAMS		
Units:	From	_____
Format: ELEC-M	To	_____
	Phas	_____
Temperature	[K]	_____
Pressure	[ATM]	_____
Vapor Frac		_____
Solid Frac		_____
Mole Flow	[KMOL/HR]	_____
Mass Flow	[KG/HR]	_____
Volume Flow	[L/MIN]	_____
Enthalpy	[MMKCAL/HR]	_____
Mass Flow	[KG/HR]	_____
H2O		_____
HNO3+		_____
H3O+		_____
NO3−		_____
OH−		_____

Figure 4.24 Aspen Plus printout [3]. Refer to flowsheet schematic, Figure 4.22, Example 4.6.

4.3 If a distillation column loses heat to the environment (non-adiabatic operation), how would that affect the separation?

4.4 If the operating line was <u>not</u> a straight line on a McCabe–Thiele plot, could you still use the methodology to determine the number of equilibrium stages? Justify your answer.

4.5 As the relative volatility decreases in value, does R_D increase or decrease for a given separation? How does that affect the heat requirement for the column?

4.6 Why is it reasonable to assume that a partial reboiler is always an equilibrium stage?

4.7   Nomenclature

For a binary system:

D	molar flowrate of liquid distillate
x_D	mole fraction of most volatile component in distillate
B	molar flowrate of liquid bottoms
x_B	mole fraction of most volatile component in bottoms
F	molar flowrate of feed
x_F	mole fraction of most volatile component in feed
V, L	molar flowrates of vapor and liquid streams in enriching section
$\overline{V}, \overline{L}$	molar flowrates of vapors and liquid streams in stripping section
n, m	number of stages in enriching and stripping sections, respectively
$y_{n,m}$	mole fraction of more volatile component in vapor at stage n, m
$x_{n,m}$	mole fraction of more volatile component in liquid at stage n, m
h_F	molar enthalpy of feed stream
Q_C	energy requirement of condenser (should be negative)
Q_R	energy requirement of reboiler (should be positive)
h_D	molar enthalpy of distillate
h_B	molar enthalpy of bottoms

4.8   Problems

4.1 Explain why a partial condenser is an equilibrium stage and a total condenser is not.

4.2 Show the derivation of the q-line equation.

4.3 For steam distillation, derive the equation for the intersection of the operating line with the diagonal ($y = x$) line and the x-axis.

4.4 An existing distillation column has five equilibrium stages, a partial reboiler and a total condenser. It is planned to use this column to produce a 75 mol% ethanol–25 mol% water distillate from a 10 mol% ethanol in water feed. The feed is a saturated liquid. Assume a basis of 100 mol/hr. Determine the operating conditions and feed stage for the column to be used for this application.

5

Extraction

For problems, sweat is a good solvent.

– ANONYMOUS

5.1 Objectives

1 Determine the number of equilibrium stages in a partially miscible system using a:
   (a) cross-flow cascade;
   (b) countercurrent cascade.
2 Determine the minimum solvent flowrate.
3 Determine the number of equilibrium stages for an extraction process with intermediate feed stages.
4 Use the McCabe–Thiele graphical method to determine the number of equilibrium stages for completely immiscible systems.

5.2 Background

Extraction is a process in which one or more solutes are removed from a liquid by transferring the solute(s) into a second liquid phase (the mass-separating agent, or MSA), for which the solutes have a higher affinity. Just as in other separations involving an MSA, the two phases are brought into intimate contact with each other, then separated. Extraction depends on differences in both solute solubility and density of the two phases. The solubility difference of the solute between the two liquid phases makes separation possible, while the difference in density allows the two liquid phases to be separated from each other. Leaching, or solid extraction, is a similar process in which solute(s) are removed from a solid by a liquid mass-separating agent for which they have an affinity. Extraction is typically analyzed as an equilibrium-limited separation.

An environmental application of liquid extraction is the removal of trace organics from water. Examples are the separation of acetic acid–water mixtures and removal of solvents, insecticides, pesticides, etc., from water. It can also be applied to the separation of liquids with close boiling points or those that form azeotropes, such that distillation is not useful. In addition, zero- or low-volatility compounds, such as metals and organometallic derivatives, can be separated by liquid extraction as can mixtures of water–hydrogen bonded compounds, such as formaldehyde. Solid extraction (leaching) can be used to remove organics or heavy metals from contaminated soils, sludges or contaminated equipment.

One advantage of extraction separations is that they can be performed at ambient temperatures. Thus, extraction is relatively energy efficient and can be applied to separations involving thermally unstable molecules. In addition, extraction processes can accommodate changes in flowrates and the solvent (mass-separating agent) can be recovered and recycled for reuse. The primary disadvantage of extraction is the complexity arising from the addition of a mass-separating agent to the system. While in distillation the simplest system is a binary, in extraction it is a ternary. The solvent must be stored, recovered, and recycled. Additionally, some of the equipment required in extraction systems is complicated and expensive. Because it is based upon solubility differences between phases, it is not possible to remove all the solute from the feed phase to the solvent phase, so pure products are not possible. Finally, models to predict efficiency and capacity in extraction are more complex than those for distillation, as is scale-up.

In general, when either distillation or extraction is feasible to achieve a separation, distillation is the method of choice. In distillation there is no mass-separating agent to be recovered. In extraction, on the other hand, the solvent is recovered continuously for reuse, usually by distillation. The addition of a new species to any system requires a separation process for its recovery. Thus, extraction separations must include two separation steps, while distillation separations require only one.

Because it is key to the success of any extraction process, the optimal solvent must meet several criteria. The distribution (partition) coefficient must be high under operating conditions. The distribution coefficient is a ratio describing the solubility of the solute in the solvent to that in the original feed stream. A high number indicates the solute has a higher affinity for the solvent than for the feed material. The solvent should have a low solubility in the feed material, such that the solvent does not replace the solute as the contaminating species. Likewise, the other feed material should not be highly soluble in the solvent and the affinity of the solvent for the solute should not be so high that recovery is prohibitive. The solvent should be compatible with both the feed stream (non-reactive) and the physical system (non-corrosive). The solvent should also have a low vapor pressure, viscosity and freezing point for easy handling and storage. In addition, the solvent should be non-toxic and non-flammable.

The solvent is recovered in a second step, typically by distillation. So, the solute–solvent pair should not form an azeotrope and the more volatile component should be the minor component in the mixture.

5.3 Environmental applications

Liquid/liquid extraction (LLE) is used for the removal of low levels of organic compounds or heavy metals from aqueous streams. It is normally evaluated relative to steam stripping or distillation. There are several factors to consider:

1 Physical properties of solute (contaminant) and water. If the boiling point of the solute is significantly below that of water, the separation is usually performed by steam stripping. One important exception is the formation of non-ideal liquid-phase solutions and minimum temperature azeotropes. This allows the solute to be removed by steam stripping even though the pure solute has a higher boiling point. So, high-boiling-point organics are good candidates for LLE. They may form azeotropic mixtures with water but, if the boiling point of the azeotrope is close to that of water, there is not an advantage for steam stripping.

   An additional mechanism for non-ideal liquid-phase solutions is hydrogen bonding. An example is formaldehyde that has a boiling point (-19 °C) much lower than that of water. But, it hydrogen bonds very strongly with water so it is difficult to remove by stripping. LLE is an effective approach for this separation.

2 Low concentrations of carboxylic acids. Acetic acid and formic acids have boiling points near (formic) or above (acetic) water. Acetic acid does not form an azeotrope and the boiling point of the formic acid–water azeotrope is 107 °C. So, LLE is a viable option to extract these acids from water.

3 Metals and organometallic derivatives. These compounds are non-volatile and thus cannot be removed by stripping. LLE is typically used to remove these compounds. A chemical complexing agent (typically an ion-exchanger) is added to the solvent phase to extract the metal. An acid is contacted with the solvent phase to strip the metal and regenerate the complexing agent.

5.4 Definition of extraction terms

- Solute (A): the species to be removed from the liquid diluent stream.
- Diluent (D): the component containing the solute which is to be removed.
- Solvent (S): the second liquid stream which will remove the solute from the diluent.
- Raffinate (R): the exiting phase which has a high concentration of diluent (and less A).
- Extract (E): the exiting phase which has a high concentration of solvent (and more A).

5.5   Extraction equipment

5.5.1   Liquid extraction

As in gas absorption and distillation, liquid extraction requires that two phases be brought into intimate contact with each other to ensure transfer of the solute from the diluent to the solvent, and then separated. In absorption and distillation there are two phases (liquid and vapor) with significantly different densities. This is not the case with extraction where both solvent and diluent solutions are liquids, often with similar densities. Because of this and the immiscibility requirement, the two liquids are often difficult to mix and even more difficult to separate. In addition, liquids have sufficiently high viscosities that they require pumps to maintain flow. Most extraction processes include mechanical energy for pumping, mixing, and separating the liquids.

Liquid extraction can be performed in a series of mixer–settler vessels as shown in Figure 5.1. The diluent stream flows countercurrent to the solvent stream and each mixer–settler pair is an equilibrium-limited stage. Then, the phases are gravity separated in a settler. A reasonable difference in densities is required between the two phases. The raffinate layer then moves to the next mixer and the extract to the next mixer in the opposite flow direction. The extract can be either more or less dense than the raffinate, so that it can be the top or bottom layer in the settler.

Spray or packed extraction towers, as shown in Figure 5.2, are similar to those used in gas absorption with the exception that two liquids, rather than a gas and a liquid, are flowing countercurrent to each other. The less dense, or lighter, liquid enters the tower bottom through a distributor, which creates small drops flowing upwards. These rise through the

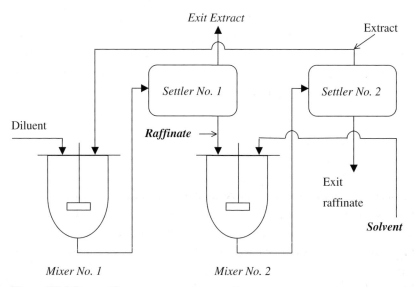

Figure 5.1 Mixer–settler extractor.

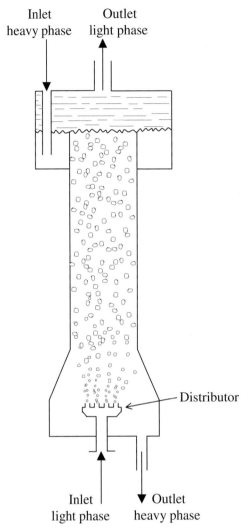

Inlet
heavy phase

Outlet
light phase

Distributor

Inlet
light phase

Outlet
heavy phase

Figure 5.2 Spray-tower extractor.

more dense, or heavy, liquid, which flows continuously downward. The drops of lighter liquid are collected at the top and the heavy liquid exits the bottom of the column. This scheme can also be reversed so that the heavy liquid is dispersed into droplets at the top of the column that fall through a continuously upward-flowing lighter fluid. The choice of which phase is dispersed depends upon mass transfer rates and which phase is the diluent or the solvent. Whichever scenario results in the highest rate of mass transfer is the best for a specific separation.

Regardless of which phase is dispersed there is a continuous transfer of solute from the diluent to the solvent. Equilibrium is never completely reached at each point in the column. Rather, the difference between equilibrium and operating conditions governs the driving force for mass transfer.

In spray towers the highest rates of mass transfer tend to occur close to the distributor plates. At small distances above and below the plate the dispersed phase tends to recoalesce and mass transfer declines significantly, such that it is more effective to add distributors to redisperse the droplets than to increase the height of the tower. Another solution is to add packing similar to that used in absorption towers. The packing causes the drops to coalesce and reform and, thus, reduces the height of each theoretical transfer unit.

The two streams flowing countercurrent creates the possibility of flooding. If the lighter phase is dispersed and the downward flowrate of the heavier phase becomes too high, the lighter-phase droplets become held up and exit with the heavy phase through the column bottom. If the heavier phase is dispersed and the flow of the lighter phase becomes too high, the heavy-phase droplets become held up on top of the lighter phase and exit the top of the column. Flooding velocity for the continuous phase as a function of that for the distributed phase can be estimated from Figure 5.3.

The x-axis is the group $\dfrac{\left(\sqrt{\overline{V}_{s,c}} + \sqrt{\overline{V}_{s,d}}\right)^2 \rho_c}{a_v \mu_c}$

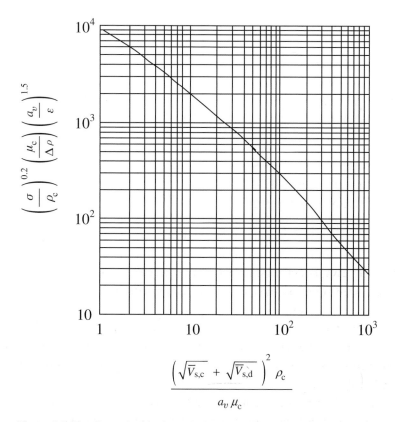

Figure 5.3 Flooding velocities in packed extraction towers [1]. L. McCabe, *Unit Operations of Chemical Engineering*. Reproduced with permission of McGraw-Hill companies.

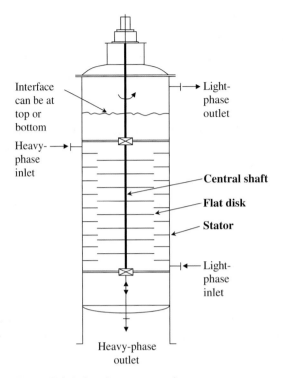

Figure 5.4 Agitated-tower extraction.

and the y-axis is the group $\left(\dfrac{\sigma}{\rho_c}\right)^{0.2} \dfrac{\mu_c}{\Delta\rho} \left(\dfrac{a_v}{\varepsilon}\right)^{1.5}$,

where $\overline{V}_{s,c}$, $\overline{V}_{s,d}$ = superficial velocities of continuous and dispersed

   phases, respectively, ft/h

μ_c = viscosity of continuous phase, lb/ft · h

σ = interfacial tension between phases, dyn/cm

ρ_c = density of continuous phase, lb/ft^3

$\Delta\rho$ = density difference between phases, lb/ft^3

a_v = specific surface area of packing, ft^2/ft^3

ε = fraction voids or porosity of packed section.

It is important to note that neither group is dimensionless, so proper units must be used.

The y-axis can be estimated from physical properties and the corresponding x-axis found. Then, the required dispersed or continuous solvent phase velocity can be found as a function of a specified diluent flowrate. If the diluent is the dispersed phase, the continuous solvent flowrate is found and likewise if the diluent is the continuous phase, the dispersed solvent flowrate is determined. Obviously, there will be some variation with column diameter and the best design is most easily determined with a spreadsheet estimation of flowrates corresponding to different diameters. Both economic and mass transfer considerations determine the "best choice" of many technical solutions.

The tower extractors described above rely purely upon density differences between the two liquids and gravity flow to achieve good mixing and separation of the phases. In some tower extractors, like the one shown in Figure 5.4, mechanical mixing is created by agitators mounted on a central rotating shaft to enhance the mass transfer. Flat disks disperse the liquids and impel them outward toward the tower walls, where *stator* rings create "quiet" zones where the phases can separate. This design gives essentially a cascade of mixer–settlers with greatly increased mixing and settling efficiencies. The disadvantage of these designs is the problem of maintaining the internal moving parts and the special construction materials that may be required for a corrosive or reactive environment.

5.6   Leaching processes

As previously stated, leaching is another extraction process in which a liquid is used to remove soluble matter from its mixture with an inert solid. With a few extra considerations, the equilibrium analysis of leaching is the same as for liquid extraction. Several assumptions are made in designing leaching processes. These can be rendered correct with the proper choice of solvent. It is assumed that the solid is insoluble in the solvent (dirt will not dissolve in water) and the flowrate of solids is essentially constant throughout the process. The solid, on the other hand, is porous and will often retain a portion of the solvent.

One way in which leaching differs is that, depending upon the solute and the solid material from which it is leached, the solid may remain the same or change considerably in form. For example, when impurities are leached from wood in the paper-making process, the final wood product retains some solvent and becomes a pulpy, mushy mixture. Coffee beans, on the other hand, are relatively unchanged when leached with hot water to make coffee. The desirable end product determines whether or not this is significant. In coffee making, the leftover ground beans are the waste product, so their final form is irrelevant. In paper making, the wood pulp is the product, so a change is important.

There are two fundamental mechanical systems in which leaching can be performed, dispersed solids leaching and stationary solid beds. The properties of the solid to be leached determine which is applicable. When the solids form an open, permeable mass throughout the leaching process, such that solvent can be reasonably percolated through an unagitated bed of solids, stationary solid bed leaching is appropriate. This would be the case when treating highly porous soils, such as grainy sands, that do not change form when exposed to the leaching fluid. If the solids are either impermeable, such as dense clays, or if they disintegrate into the solvent, the solids must then be dispersed into the solvent and then later recovered from it. Under these conditions, dispersed solids leaching is suitable.

Dispersed solids leaching is performed in a continuous mode in which the solids are first dispersed into solvent by mechanical agitation. This mixed solids stream is then run countercurrent to the solvent stream in a series of mixer–settlers as shown in Figure 5.1. The design and estimation of the number of equilibrium-limited stages is similar to that of liquid extraction, except that one additional stage is needed to disperse the solids in the solvent. Solute concentrations will change significantly in this stage due to dilution with solvent. An illustrative example is provided in a subsequent section of this chapter. At the end of the process the leached solids residue is separated from the mixed solution by settling or filtration. The solute can be recovered from the exiting liquid solvent stream by evaporation or crystallization.

Stationary solid bed leaching is performed in a series of mixer–settlers similar to those shown in Figure 5.1. This system is best for permeable solids, but can also be used to treat reasonably semi-permeable solids if pressure is applied to force the solvent through the beds. Each mixer–settler unit is a single tank with a perforated false bottom, which both supports the solids and allows drainage of the solvent. The process is performed in batch mode. Solids are loaded into each leaching tank, sprayed with solvent until the solute concentration is adequately low, then excavated back out of the tank. Because of the downtime required to pack and excavate the tanks, more than one is required in series in a continuous process. Each vessel is analogous to an equilibrium-limited stage. The solvent then passes through each vessel in series beginning with the tank containing the solid with the lowest solute concentration (most completely extracted) and ending with the least extracted tank. The solid in any one tank is stationary until it is completely leached. The solvent system piping is arranged such that fresh solvent can be introduced to any tank and spent solvent can be withdrawn from any tank. Countercurrent flow is maintained by advancing the inlet and outlet tanks one at a time.

Another type of stationary bed extractor is shown in Figure 5.5. This Hildebrand extractor consists of a U-shaped screw conveyor with a separate auger, or screw, mechanism in each section to transport the solids through the U. Each auger rotates at a different speed to control compaction of the solids. Solids are fed into one end of the U and solvent into the other to provide countercurrent flow.

5.6.1 Partially miscible extraction

The analysis of extraction is based on the two primary modes of contact: cross-flow and countercurrent.

Cross-flow

Figure 5.6 is a schematic of a cross-flow cascade.
<u>Mass balance</u>:

$$O_0 + V_0 = M_1 = O_1 + V_1 \quad \text{(Stage 1).} \tag{5.1}$$

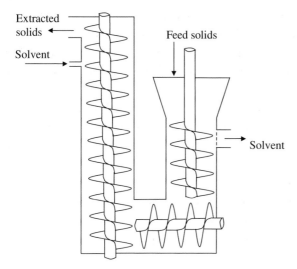

Figure 5.5 Hildebrand extractor.

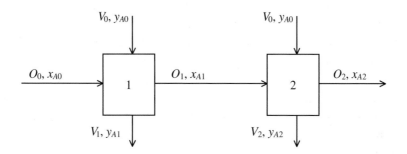

O = diluent flowrate
V = solvent flowrate
x_A, y_A = solute mass fraction in corresponding phase

Figure 5.6 Cross-flow cascade.

Therefore, from the lever-arm rule, $\overline{O_0 M_1 V_0}$ and $\overline{O_1 M_1 V_1}$ are both straight lines on a triangular composition diagram. M_1 is the mixing point for Stage 1. Since the stage is an equilibrium-limited stage, O_1 and V_1 are in equilibrium. Therefore, $\overline{O_1 M_1 V_1}$ is a tie-line.

$$O_1 + V_0 = M_2 = O_2 + V_2 \quad \text{(Stage 2),} \tag{5.2}$$

and by the same analysis as above, $\overline{O_1 M_2 V_0}$ and $\overline{O_2 M_2 V_2}$ are straight lines, and $\overline{O_2 M_2 V_2}$ is also a tie-line.

Graphically, this could be pictured in a triangular diagram (Figure 5.7). M_1 and M_2 are located using stream flowrate ratios. The dashed lines represent the tie-lines.

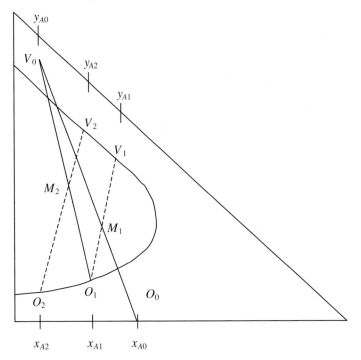

Figure 5.7 Triangular diagram depicting cross-flow cascade.

Example 5.1: cross-flow cascade

Problem:

Acetic acid (vinegar) is a common component in the waste stream of various fermentation processes. This process represents a significant environmental concern because the manufacture of pharmaceuticals can result in carboxylic acid-containing by-products. The carboxylic acid byproducts must be removed from the pharmaceutical products as well as any water streams prior to discharge. Isopropyl ether has been used as a solvent for extraction. The isopropyl ether is then distilled and recycled back to the extraction process. As an example, a feed that is 30 wt% acetic acid and 70 wt% water is fed to a two-stage cross-flow cascade. Feed flowrate is 1000 kg/hr, and 1250 kg/hr of solvent containing 99 wt% isopropyl ether and 1% acetic acid is added to each stage. Operation is at 20 °C and 1 atm where the equilibrium data are given in Table 5.1 [2]. Determine:

(a) the weight fractions of the outlet raffinate and the two outlet extract streams;

(b) the flowrate of the outlet raffinate.

Solution:

Figure 5.8 is a schematic of the example. Point O_0 and point V_0 are plotted on Figure 5.9, with a line connecting them. The *x*-coordinate of the first mixing point, M_1, was found from the equation:

$$z_{A1} = \frac{V_0 y_{A0} + O_0 x_{A0}}{V_0 + O_0} = 0.31,$$

and it was plotted as a point on the line $\overline{O_0 V_0}$ (the lever-arm rule could also have been used). A tie-line was then drawn through this mixing point; the ends of the tie-line are the points O_1 (0.73, 0.25) and V_1 (0.02, 0.06), and their compositions were read from the graph. Two mass balances (overall and one component) solved simultaneously give the flowrates of the streams exiting the first stage.

<u>Overall mass balance:</u>

$$O_1 + V_1 = 1000 \text{ kg/hr} + 1250 \text{ kg/hr} = 2250 \text{ kg/hr}.$$

<u>For water:</u>

$$0.7(1000 \text{ kg/hr}) = 0.73 O_1 + 0.02 V_1; \qquad V_1 = 1327 \text{ kg/hr}, O_1 = 923 \text{ kg/hr}.$$

In Stage 2, the stream O_1 is mixed with a solvent stream identical in composition and mass as the one in Stage 1, so the next line drawn connects O_1 with V_0. The next mixing point, M_2 is found, and the analysis for Stage 1 is repeated for Stage 2. Note that the O streams would not all be connected to V_0 if different compositions of the solvent phase were used for each stage. The compositions of O_2, V_1, and V_2 can be read directly from Figure 5.9. Expressed as fractions, they are:

$$x_{A2} = 0.78 \qquad x_{B2} = 0.205$$

$$y_{A1} = 0.02 \qquad y_{B1} = 0.06$$

$$y_{A2} = 0.015 \qquad y_{B2} = 0.045.$$

The flowrate, O_2, can be calculated using the lever-arm rule:

$$\frac{\overline{M_2 V_2}}{\overline{O_2 V_2}} \times 2250 = 837 \text{ kg/hr} = O_2.$$

Table 5.1 *Equilibrium data for Example 5.1.*

Water layer, wt%			Isopropyl ether layer, wt%		
Isopropyl ether	Water	Acetic acid	Isopropyl ether	Water	Acetic acid
1.2	98.1	0.69	99.3	0.5	0.18
1.5	97.1	1.41	98.9	0.7	0.37
1.6	95.5	2.89	98.4	0.8	0.79
1.9	91.7	6.42	97.1	1.0	1.93
2.3	84.4	13.30	93.3	1.9	4.82
3.4	71.1	25.50	84.7	3.9	11.40
4.4	58.9	36.70	71.5	6.9	21.60
10.6	45.1	44.30	58.1	10.8	31.10
16.5	37.1	46.40	48.7	15.1	36.20

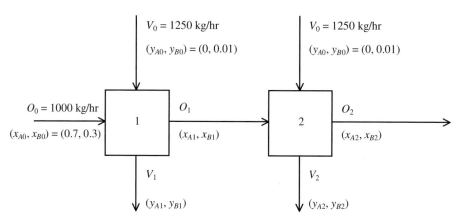

Figure 5.8 Schematic for water–acetic acid–isopropyl ether, Example 5.1.

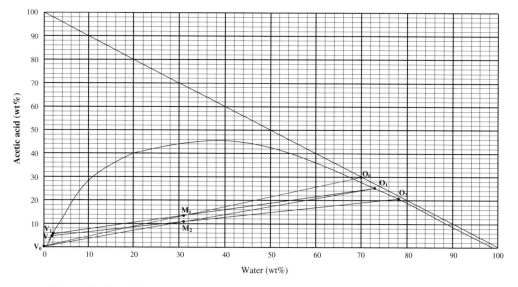

Figure 5.9 Cross-flow cascade: water-acetic acid–isopropyl ether, Example 5.1.

Countercurrent cascade (Figure 5.10)

A countercurrent cascade allows for more complete removal of the solute, and the solvent is reused from stage to stage so that less is needed.

Mass balance:

$$O_1 + V_1 = O_2 + V_0 \quad \text{(Stage 1)}$$
$$\therefore V_0 - O_1 = V_1 - O_2 = \Delta; \tag{5.3}$$

$$O_2 + V_2 = O_3 + V_1 \quad \text{(Stage 2)}$$
$$\therefore V_1 - O_2 = V_2 - O_3 = \Delta. \tag{5.4}$$

So Δ represents the difference in flowrates of passing streams between stages. This "delta

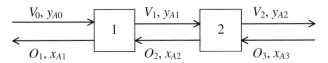

Figure 5.10 Countercurrent cascade.

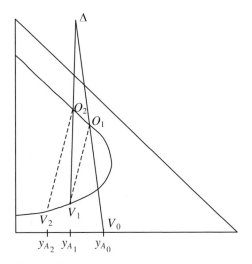

Figure 5.11 Triangular diagram for countercurrent cascade.

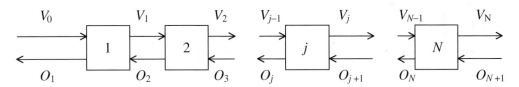

Figure 5.12 Countercurrent cascade of N stages.

point" came directly from the mass balances. Therefore, we can write:

$$V_0 = \Delta + O_1 \Rightarrow \overline{\Delta V_0 O_1} \quad \text{is a straight line and the composition of } V_0$$
$$\text{will lie between } \Delta \text{ and } O_1 \text{ (if } \Delta \text{ is positive).}$$

The same can be said for: $V_1 = \Delta + O_2$ and $V_2 = \Delta + O_3$.

The compositions corresponding to O_1 and V_1 (as well as O_2 and V_2) will lie on a tie-line since they are outlet streams from a stage and they are assumed to be in equilibrium with each other. O_3 and V_0 are not on a tie-line. This is shown graphically in Figure 5.11 (dashed lines are tie-lines).

A more general case consists of N stages (Figure 5.12).

Mass balance on system:

$$\text{In} = \text{Out} \quad \Rightarrow \quad O_{N+1} + V_0 = O_1 + V_N = M. \tag{5.5}$$

We know that M must lie on a line connecting O_1 and V_N, and O_{N+1} and V_0 (these lines are <u>not</u> tie-lines). Using the lever-arm rule, we can locate M. In other words, we can write:

$$M = O_1 + V_N \qquad M \text{ is on this lever arm;} \tag{5.6}$$

$$M = O_{N+1} + V_0 \qquad M \text{ is also on this lever arm.} \tag{5.7}$$

Therefore, we can use the lines $\overline{O_{N+1}MV_0}$ and $\overline{O_1MV_N}$ to get the values of the compositions of the streams exiting/entering the column.

[Note: O_{N+1} and V_0 are <u>not</u> on the solubility envelope because they are entering streams and are not necessarily in equilibrium. O_1 and V_N, however, <u>are</u> on the solubility envelope since they are exit streams and leave the column in equilibrium with each other.]

Mass balance on Stage 1:

$$V_0 + O_2 = O_1 + V_1$$
$$O_1 - V_0 = O_2 - V_1 = \Delta. \tag{5.8}$$

Mass balance on Stages 1 and 2:

$$V_0 + O_3 = O_1 + V_2$$
$$O_1 - V_0 = O_3 - V_2 = \Delta. \tag{5.9}$$

Mass balance on Stage j:

$$O_1 - V_0 = O_{j+1} - V_j = \Delta. \tag{5.10}$$

Mass balance on N stages (entire column):

$$O_1 - V_0 = O_{N+1} - V_N = \Delta. \tag{5.11}$$

Therefore, net flow Δ is constant through cascade.

These mass balance equations show that Δ must lie on the line $\overline{\Delta V_0 O_1}$ and also on the line $\overline{\Delta V_N O_{N+1}}$. The procedure for plotting the Δ point is illustrated in Figure 5.13. [Remember that $V_0/O_1 = \dfrac{O_1\Delta}{V_0\Delta}$. In the previous figure, $V_0 > O_1$ and Δ is positive, so that the net flow is toward the rich end of the column, i.e., the extract (V) phase.]

Now let V_N, V_0, and O_{N+1} stay the same while O_1 (raffinate) moves up and to the left. This corresponds to a "less pure" raffinate. As O_1 is allowed to move up and to the left, the ratio V_0/O_1 increases since Δ goes further to the left of the graph. When the lines $\overline{O_1 V_0 \Delta}$ and $\overline{O_{N+1} V_N \Delta}$ are parallel, $V_0/O_1 = \infty/\infty = 1$ and $\Delta = 0$.

When O_1 is allowed to move even further to the left, Δ will be to the right of the graph instead of to the left of it, and $O_1 > V_0$, and Δ is negative. This means that the net flow would be toward the lean end of the cascade, i.e., the raffinate (O) phase.

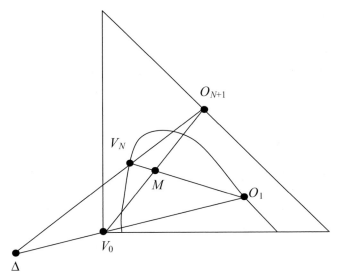

Figure 5.13 To plot the delta point: 1 Locate the points corresponding to the compositions of V_0 (solvent), $\overline{O_{N+1}MV_0}$ (feed), and O_1 (exit raffinate). 2 Find point M from a component balance or the lever-arm rule. 3 Make line $\overline{O_{N+1}MV_0}$. Extend line $\overline{O_1M}$ to the saturation curve (this is point V_N). 4 Extend $\overline{O_{N+1}V_N}$ and $\overline{O_1V_0}$; the intersection is at Δ.

Example 5.2: countercurrent extraction

Problem:

A 1000 kg/hr stream of a solution containing 30 wt% acetic acid and 70 wt% water is to be fed to a countercurrent extraction process. The solvent is 99% isopropyl ether and 1% acetic acid, and has an inlet flowrate of 2500 kg/hr. The exiting raffinate stream should contain 10 wt% acetic acid. The equilibrium data are the same as given in Table 5.1 for the cross-flow Example 5.1. Find the number of equilibrium-limited stages required.

Solution:

Known: $O_{N+1} = 1000$, $y_{AN+1} = 0.70$, $y_{BN+1} = 0.30$

$\qquad\qquad V_0 = 2500$, $x_{A0} = 0$, $x_{B0} = 0$

$\qquad\qquad O_1 = ?$, $y_{B1} = 0.10$ (and it's on the saturation curve).

Method (see Figure 5.14):

1 Plot equilibrium data from previous example (plotting the conjugate line will be really helpful for drawing tie-lines in this case).

2 Plot the line $\overline{V_0O_{N+1}}$ and the point O_1.

3 Find the mixing point:

$$z_{BM} = \frac{2500(0.01) + 1000(0.30)}{2500 + 1000} = 0.14 \text{ (and it's on the line } \overline{V_0O_{N+1}}).$$

4 Find the line through $\overline{O_1 M}$ to get V_N (it's on the saturation curve).

5 Extend the lines $\overline{V_0 O_1}$ and $\overline{V_N O_{N+1}}$: the intersection is the Δ point.

6 Stepping off stages:
  – Use a tie-line from O_1 to find V_1.
  – Draw and extend $\overline{\Delta V_1}$. It crosses the solubility curve at O_2.
  – Use the tie-line from O_2 to find V_2, etc.

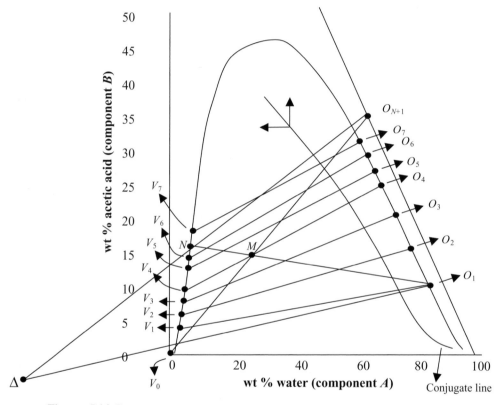

Figure 5.14 Countercurrent cascade: water–acetic acid–isopropyl ether, Example 5.2.

5.7 Minimum solvent flowrate

Remember that:

$$\frac{V_0}{O_{N+1}} = \frac{\text{solvent flowrate}}{\text{feed flowrate}}. \qquad (5.12)$$

Note that in Figure 5.13, this ratio decreases as the point M moves toward the point V_0 (lever-arm rule). This causes a decrease in the solute concentration of V_N. When the line $\overline{\Delta V_0 O_{N+1}}$ changes such that it falls exactly on a tie-line (equilibrium and operating lines

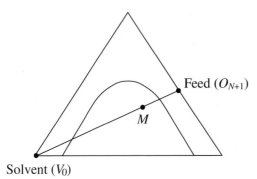

Figure 5.15

are intersecting), a pinch point is reached. If you try to step off stages at a pinch point, you will see that the separation will never improve even with infinite stages. This is where the minimum ratio O_{N+1}/V_0 is located, which can be used to find the minimum amount of solvent required to obtain the desired separation. If less solvent is used, the desired separation is impossible; while if more solvent is used, the separation can be achieved with a finite number of stages. Note that this concept is analogous to minimum reflux (distillation).

To get the minimum solvent/feed flowrate, we want to find the mixing point M (Figure 5.15) that:

(a) lies in the two-phase region;

(b) minimizes the line length $\overline{MO_{N+1}}$ in comparison to $\overline{MV_0}$. From the lever-arm rule,

$$\frac{\overline{MO_{N+1}}}{\overline{MV_0}} = \frac{S}{F};\tag{5.13}$$

(c) is consistent with outlet requirements (compositions) of raffinate (R) or extract (E) streams from the cascade. Since this is based on mass balances, this is what is needed irrespective of which direction the tie-lines (equilibrium lines) slope in the two-phase region;

(d) gives the largest value of V_0/O_{N+1} that still would cause a pinch point.

5.7.1 Case 1: Tie-lines slope downward to the left on the equilibrium diagram (Figure 5.16).

Using either Δ_1 or Δ_2 as the Δ_{min} will result in a pinch point since both were found by extending tie-lines (dashed lines). Δ_1 is the extension of the tie-line that would pass through O_{N+1}. Using Figure 5.16, draw the line $\overline{V_0 O_{N+1}}$.

(a) For point Δ_1 as the Δ_{min} point, the operating line is $\overline{\Delta_1 O_{N+1}}$. V' is the composition of the extract phase. The superscript ($'$) is to relate it to Δ_1. Draw $\overline{V'O_1}$. The intersection of $\overline{V'O_1}$ and $\overline{V_0 O_{N+1}}$ is M_1, the mixing point. The ratio $\overline{M_1 O_{N+1}}/M_1 V_0 = V_0/O_{N+1}$.

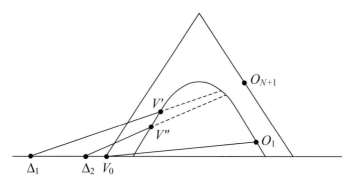

Figure 5.16

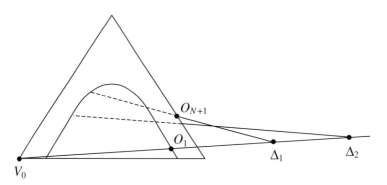

Figure 5.17

(b) For point Δ_2 as the $\Delta_{\min}$ point, the operating line is $\overline{\Delta_2 O_{N+1}}$. Repeat procedure to find V_0/O_{N+1}. You will see that in this case, V_0/O_{N+1} is <u>larger.</u>

As the $\Delta_{\min}$ point moves toward point V_0, the ratio V_0/O_{N+1} will <u>increase</u> but it is still the <u>minimum</u> that we use. Since the actual V_0/O_{N+1} is a multiple of the minimum, we need to make sure that the actual value will not cause a pinch point anywhere in the column.

5.7.2 Case 2: Tie-lines slope downward to the right on the equilibrium diagram (Figure 5.17)

Again, Δ_1 and Δ_2 correspond to two pinch points. Follow the same sequence as for Case 1 to find the mixing points corresponding to Δ_1 and Δ_2.

The ratio V_0/O_{N+1} for Δ_2 is larger than the ratio obtained using Δ_1. As $\Delta_{\min}$ point moves away from point V_0, the ratio V_0/O_{N+1} will <u>increase</u> but it is still the minimum that we use to insure that the actual value that we use does not cause a pinch point in the column.

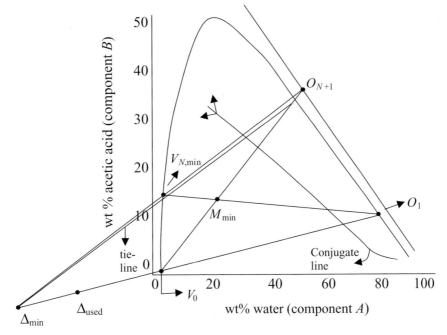

Figure 5.18 Minimum solvent flowrate calculation: water–acetic acid–isopropyl ether, Example 5.3.

Example 5.3: minimum solvent flowrate

Problem:

Find the minimum solvent flowrate for Example 5.2.

Solution:

The previous steps were used to create Figure 5.18.

$$\left(\frac{V_0}{O_{N+1}}\right)_{min} = \frac{\overline{O_{N+1}M_{min}}}{\overline{V_0 M_{min}}} = 1.6.$$

Since $O_{N+1} = 1000$ kg/hr, $V_{0min} = 1600$ kg/hr. The actual solvent flowrate in the previous example was 2500 kg/hr, so $V_0/V_{0min} = 1.6$ for that extraction process. Using even more solvent would decrease the required number of equilibrium stages.

5.8   Countercurrent extraction with feed at intermediate stage

The concentration of solute (A) in the outlet extract stream is limited to a relatively low value because the outlet extract stream is a passing stream with feed O_{N+1}. So the maximum concentration of A in V_N will occur when the column is operated at the minimum solvent flowrate. As with distillation, this limitation can be overcome by using

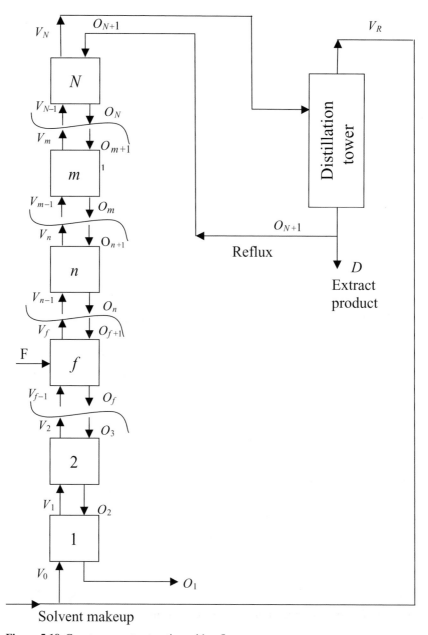

Figure 5.19 Countercurrent extraction with reflux.

extract reflux. Extract reflux, as shown in Figure 5.19, is used most often for systems where the solute is only partially miscible with the extracting solvent.

Extract reflux involves taking part of the exiting extract stream and changing it into a raffinate phase by removing solvent and returning this raffinate phase to the column. The solvent is usually removed by distillation, but can be removed by other methods such as flash distillation, reverse osmosis, ultrafiltration, or another extraction column.

In Figure 5.19,

O_{N+1} is the reflux stream

and

$$\frac{O_{N+1}}{D} = \text{reflux ratio.}$$

The analysis on the upper and lower sections is not dependent on the method used to recover the solvent phase.

Mass balance on upper (enriching) section:

[Remember: the enriching section corresponds to increasing concentrations of solute.]

$$V_N - O_{N+1} = V_{N-1} - O_N = \Delta'$$
$$V_{n-1} - O_n = \Delta'. \tag{5.14}$$

So the point Δ' is on the line $\overline{O_{N+1} V_N \Delta'}$. If the composition of O_{N+1} (which has the same composition as the extract product) and V_N are known as well as the total flowrates, one can locate Δ', which is then used to step off stages for the enriching section.

Mass balance on lower (stripping) section:

$$V_{f-1} - O_f = \Delta \tag{5.15}$$
$$V_1 - O_2 = \Delta$$
$$V_0 - O_1 = \Delta.$$

So the point Δ is on the line $\overline{V_0 O_1 \Delta}$. If the compositions and total flowrates of V_0 and O_1 are known, Δ can be located.

Overall mass balance on the column:

$$F + V_0 - O_1 = V_N - O_{N+1} \quad \text{where } F \text{ is the feed}$$
$$F + \Delta = \Delta'. \tag{5.16}$$

Therefore, $\overline{F \Delta \Delta'}$ *are on the same line.*

Procedure for solving extraction with reflux problems

1 Plot equilibrium data.
2 Plot the concentrations of the known variables: V_0, F, D (same as O_{N+1}), O_1 (O_1 is on the phase envelope).
3 Find V_N from mass balances around solvent separator (it is a saturated stream).
4 Find Δ' using mass balances and the internal reflux ratio.
5 Locate feed position.

The optimum feed location is where you switch from using one Δ point to another (i.e., point F), see Figure 5.20.

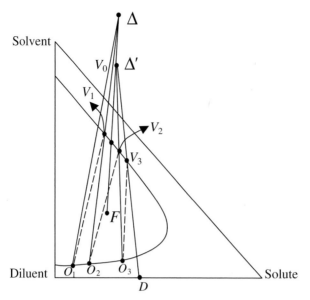

Figure 5.20 Countercurrent extraction with reflux.

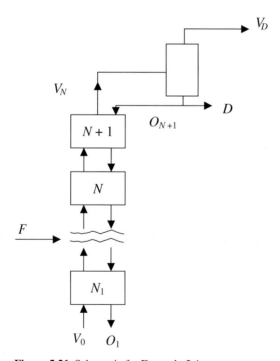

Figure 5.21 Schematic for Example 5.4.

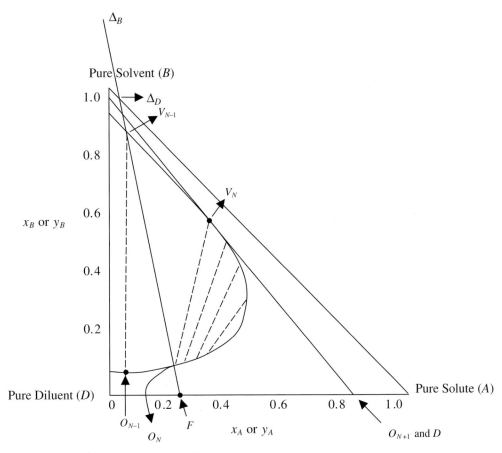

Figure 5.22 Phase diagram for Example 5.4.

5.9 Minimum and total reflux

Minimum reflux corresponds to the overlap of an operating line and a tie-line (infinite stages at a pinch point). This concept is similar to minimum solvent flowrate for an extraction process without reflux. Total reflux corresponds to the minimum number of stages. Remember that total reflux means that no streams are going into or out of the column, so that F, B, and D are zero, and $\Delta = \Delta'$.

Example 5.4: countercurrent separation with reflux

Problem:
Refer to Figures 5.21 and 5.22. Given:

$$V_0/F = 0.4; \qquad O_{N+1}/D = 3.$$

On the phase diagram (Figure 5.22):

$$x_{AF} = 0.25$$

$$x_{A1} = 0.06$$

$$y_0 = 0.0$$

$v_D =$ is pure solvent; $Y_{AD} = 0.0$

$x_{AD} = 0.82$, same concentration as the V_1 stream (solvent-free basis) since V_D is pure solvent.

Determine (a) the number of equilibrium stages; and (b) location of the feed stage.

Solution:

Mass balance on reflux distillation column:

$$V_N = V_D + O_{N+1} + D.$$

Since V_D is pure B and O_{N+1} and D have no solvent present, the line connecting pure B and O_{N+1} and D must contain V_N, which must also be on the solubility envelope since it is an outlet equilibrium stream. Therefore, V_N corresponds to $y_{AN} = 0.33$ from Figure 5.22.

$$\frac{O_{N1} + D}{V_N} = \frac{\overline{BV_N}}{\overline{BO_{N+1}}} = \frac{0.33}{0.82} = 0.4 \quad \text{(measuring horizontal distances)}$$

$$\frac{1 + (D/O_{N+1})}{(V_N/O_{N+1})} = 0.4.$$

Rearranging:

$$\frac{O_{N+1}}{V_N} = \frac{0.4}{1 + \frac{1}{3}} = 0.3.$$

Mass balance on upper section:

$$\Delta_D = V_N - O_{N+1}.$$

Therefore, Δ_D is on line through V_N and O_{N+1}.

$$\frac{\overline{\Delta_D V_N}}{\overline{\Delta_D O_{N+1}}} = \frac{O_{N+1}}{V_N} = 0.3$$

$$= \frac{x}{x + (0.82 - 0.33)}, \quad \text{again measuring horizontal distances,}$$

so $x = 0.21 =$ horizontal distance to left of V_N for point Δ_D.

Mass balance on total column:

$$F + \Delta_B = \Delta_D$$

$$\Delta_B = V_0 - O_1$$

$\therefore \Delta_B$ is on line through F and Δ_D and also on line through V_0 and O_1.

Hence, (a) two equilibrium stages are necessary for separation.

(b) Feed plate is second plate since we do not switch to Δ_B.

No rectifying section needed.

5.10   Immiscible extraction: McCabe–Thiele analysis

Some extraction systems are such that the solvent and diluent phases are almost completely immiscible in each other. Hence, separation yields an extract phase essentially free of diluent and a raffinate phase that is almost pure diluent. This greatly simplifies the characterization of the system. When partial miscibility for an extraction process is very low, the system may be considered immiscible and application of McCabe–Thiele analysis is appropriate. It is important to note that McCabe–Thiele analysis for immiscible extraction applies to a countercurrent cascade. The McCabe–Thiele analysis for immiscible extraction is analogous to the analysis for absorption and stripping processes. Consider the flow scheme shown in Figure 5.23,

where F_D = mass flowrate of diluent (feed)
  X_j = weight ratio of solute in diluent leaving stage j (kg A/kg D)
  E_j = mass flowrate of extract (spent solvent phase) leaving stage j
  Y_j = weight ratio of solute in solvent leaving stage j (kg A/kg S)
  R_j = mass flowrate of raffinate (purified product) leaving stage j
  F_S = mass flowrate of solvent.

The assumption that the diluent and the solvent are totally immiscible means that their flowrates (F_D and F_S) are constant, so that the weight ratios can be found from weight fractions:

$$X = \frac{x}{1-x} \quad \text{and} \quad Y = \frac{y}{1-y} \quad [\textit{only true for immiscible systems!}] \quad (5.17)$$

where X is kg solute/kg diluent and Y is kg solute/kg solvent.

The notation here may be confusing, because there is no vapor phase involved. The difference between x and y (or X and Y) is that x's are used to describe the amount of the solute in the

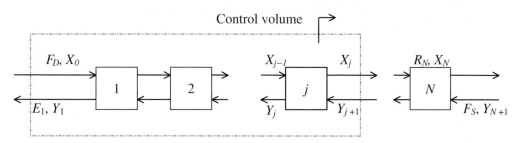

Figure 5.23 Countercurrent cascade schematic.

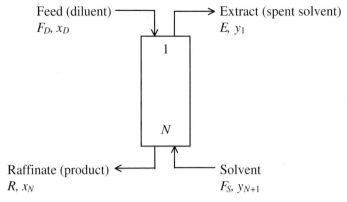

Figure 5.24 Schematic for water purification, Example 5.5.

diluent phase, while y's are used to describe the amount of the same solute in the solvent phase.

<u>Mass balance around control volume:</u>

$$F_S Y_1 + F_D X_j = F_S Y_{j+1} + F_D X_0 \Rightarrow Y_{j+1} = \frac{F_D}{F_S} X_j + \left(Y_1 - \frac{F_D}{F_S} X_0 \right).$$ (5.18)

Since F_D and F_S are constant for an immiscible system, the operating line is a straight line.

<u>Equilibrium data:</u>

Equilibrium data for dilute extractions are usually given in terms of the distribution ratio,

$$K_d = \frac{Y_A}{X_A}$$ (5.19)

in weight or mole fractions. K_d is usually constant for very dilute systems, but can become a function of concentration. It is also temperature and pH dependent.

Example 5.5: countercurrent immiscible extraction

Problem:

Figure 5.24 is a schematic of a water purification system. A 20 wt% mixture of acetic acid in water is to be extracted with 1-butanol. The outlet concentrations of acetic acid should be 5 wt% in the water phase and 10 wt% in the 1-butanol phase. Pure solvent is used. Find the number, N, of equilibrium stages required and the ratio, F_D/F_S, of water to 1-butanol. The distribution coefficient, K_d, for this system is 1.6.

Solution:

Given:

$K_d = 1.6$

$x_D = 0.20$

$y_1 = 0.10$

$x_N = 0.05$

$y_{N+1} = 0$ [note that the solvent stream doesn't always have to be pure solvent].

Equilibrium line:

From the given K_d, the equilibrium relationship is $y = 1.6x$. The equilibrium line was plotted (Figure 5.25) as Y vs X, using the relationship between weight fractions (x, y) and weight ratios (X, Y) given in Equation (5.17).

Operating line:

Since points on the operating line are passing streams, we can plot the operating line using the given information about the passing streams above Stage 1 and below Stage N. These points are (X_D, Y_1) and (X_N, Y_{N+1}); remember to convert to ratios.

Stepping off stages (see Figure 5.25):

(a) Two equilibrium stages are required.

(b) The ratio of water to 1-butanol is:

$$\frac{Y_1 - Y_{N+1}}{X_0 - X_N} = \frac{0.106 - 0}{0.13 - 0.005} = 0.85;$$

this is the slope of the operating line: F_D/F_S.

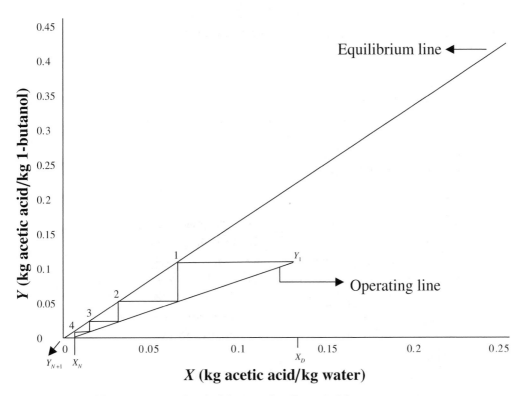

Figure 5.25 Countercurrent immiscible extraction, Example 5.5.

5.11   More extraction-related examples

Example 5.6: countercurrent vs cross-flow extraction

Problem:

Suppose we have a feed stream which contains acetic acid (A) and water (W). Because it would not be economically feasible to incinerate the water stream, we'd like to extract the acetic acid into an isopropyl ether (I) phase. The isopropyl ether–acetic acid stream could then be used as a fuel stream to another process. Note that the isopropyl ether has a limited solubility in the water, and vice versa.

Graphically analyze two configurations of equilibrium stages (Figure 5.26): countercurrent and cross-flow. A small number of mass balance calculations are required in addition to the graphical work. All compositions given are in mass fractions. Use two equilibrium stages in each of the configurations.

Solution:

Given:

$$V_3 = 30 \text{ lb/hr} \qquad V_3/2 = 15 \text{ lb/hr} \qquad O_0 = 10 \text{ lb/hr}$$
$$y_3(A) = 0 \qquad\qquad y_3(I) = 1.0 \qquad\quad\ y_3(W) = 0$$
$$x_0(A) = 0.45 \qquad\quad x_0(I) = 0 \qquad\qquad x_0(W) = 0.55.$$

Equilibrium data can be found in Table 5.1, Example 5.1.

Additional information valid for countercurrent *only*:

$$y_1(A) = 0.125 \quad x_2(A) = 0.13.$$

The solution for the countercurrent cascade (Figure 5.27) is slightly different than in Example 5.5, because now the number of stages instead of the product concentration is specified. From the information given, the line $\frac{V_3}{2}O$ can be drawn.

V_1 and O_2 can also be plotted from the given information because they both lie on the solubility envelope (they are leaving the column in equilibrium). The intersection of the lines $\overline{O_0 V_1}$ and $\frac{V_3}{2} O_2$ locates the Δ point.

The mixing point is also shown on the graph, but it is not used to locate V_1 as usual since V_1 is already specified. A tie-line through V_1 locates the point O_1, and the line $\overline{\Delta O_1}$ crosses the solubility envelope at the point V_2.

Now all of the streams are located, and their compositions can be read from the graph. Using the lever-arm rule to find the amounts of the streams:

$$\frac{V_1}{O_0} = \frac{\overline{O_0 \Delta}}{\overline{V_1 \Delta}} = \frac{14.75 \text{ cm}}{4.7 \text{ cm}} = 3.14 \text{ (using a ruler).} \qquad \therefore V_1 = 31.4 \text{ lb/hr.}$$

$$\frac{O_1}{V_2} = \frac{\overline{\Delta V_2}}{\overline{\Delta O_1}} = \frac{3.15 \text{ cm}}{12.85 \text{ cm}} = 0.25, \quad \text{also:} \quad O_1 + V_1 = O_0 + V_2.$$

Solving simultaneously gives $V_2 = 28.5 \text{ lb/hr}$ and $O_1 = 7.1 \text{ lb/hr}$.

Balances on each of the three components (Table 5.2) around the entire system are in error by between 0.5 and 3.5 lbs. This is much worse than the cross-flow case (Figure 5.28), probably because the results of subsequent stages depend on the results of the previous stage (i.e., errors propagate).

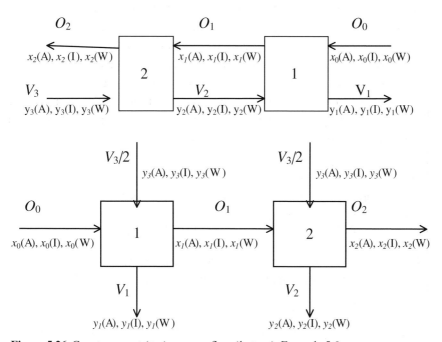

Figure 5.26 Countercurrent (*top*) vs cross-flow (*bottom*), Example 5.6.

Table 5.2 *Balances on each of the three components, Example 5.6.*

	Countercurrent	Cross-flow
V_1	31.4 lb/hr	18.67 lb/hr
$y_1(A), y_1(W), y_1(I)$	0.125, 0.040, 0.835	0.140, 0.055, 0.805
V_2	28.5 lb/hr	16.79 lb/hr
$y_2(A), y_2(W), y_2(I)$	0.043, 0.020, 0.937	0.065, 0.025, 0.910
O_1	7.1 lb/hr	6.33 lb/hr
$x_1(A), x_1(I), x_1(W)$	0.270, 0.690, 0.040	0.298, 0.670, 0.032
O_2	8.6 lb/hr	4.54 lb/hr
$x_2(A), x_2(I), x_2(W)$	0.130, 0.850, 0.020	0.175, 0.800, 0.025

Note that the countercurrent system provides a higher product flowrate as well as improved separation.

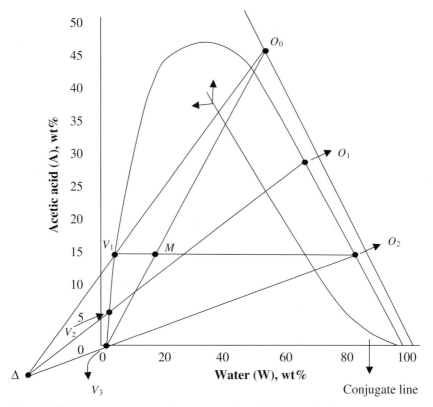

Figure 5.27 Countercurrent cascade: water–acetic acid–isopropyl ether, Example 5.6.

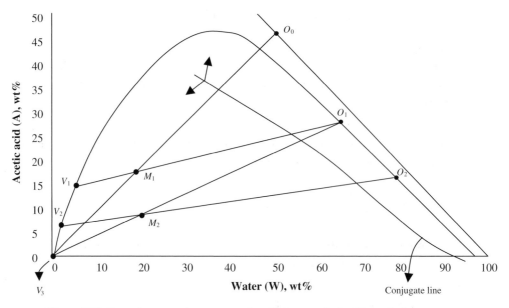

Figure 5.28 Cross-flow cascade: water–acetic acid–isopropyl ether, Example 5.6.

Example 5.7

Problem:

100 kg/hr of waxed paper is to be dewaxed in a continuous countercurrent leaching process. The paper contains 25 wt% wax and 75 wt% paper pulp. The paper pulp should contain less than 0.3 kg of wax per 100 kg paper pulp. Entering solvent contains 0.045 kg of wax per 100 kg solvent. The paper pulp retains 1.5 kg solvent for every kg of entering diluent.

Extract should contain 5 kg wax for every 100 kg solvent. How many equilibrium-limited stages are required?

Solution:

Begin with a mass balance. Assume that all of the solvent retained by the paper pulp is done so in the first contact stage and, further, that no separation occurs in this stage, only mixing and solvent retention.

Figure 5.29 is a schematic flow diagram for the solvent. Since 1.5 kg solvent are retained for every 1 kg diluent, we can say that 100 kg diluent + 150 kg solvent exit Stage 1 for a total of 250 kg. Wax and paper pulp are then diluted as follows:
25 kg/hr wax enters Stage 1

$$\text{Leaving Stage 1: } x = \frac{25\,\text{kg/hr}}{250\,\text{kg/hr}} = 0.1.$$

Next, a mass balance gives us S, the solvent flowrate:

Wax in: $100\,\text{kg/hr}\,(0.25) + 4.5 \times 10^{-4}\,S$
Wax out: $100\,\text{kg/hr}\,(3.0 \times 10^{-3}) + (S - 150)\,\text{kg/hr}\,(0.05).$

Solving, $S = 650$ kg/hr.

Figure 5.30 is a redrawn schematic diagram for the solvent. In the figure, the *denote concentrations in equilibrium, i.e., raffinate concentration in equilibrium with solvent concentrations.

We need to calculate x_a, the exiting wax concentration in the solvent prior to Stage 1. Since we assume no separation occurs in Stage 1, we can estimate it as a function of solvent flowrates and wax concentration leaving in the extract

$$x_a = \frac{(0.05)(500)}{650} = 3.9 \times 10^{-2}.$$

We must do the same to get x_b^*. Problem statement says 0.3 kg wax per 100 kg paper pulp. Total stream flow is 250 kg/hr. So,

$$x_b^* = 0.003 \left(\frac{100}{250} \right) = 1.2 \times 10^{-3}.$$

After Stage 1, we can assume constant flowrates. Hence, the Kremser equation, Equation (3.51), can be applied:

$$N = \ln\left(\frac{x_b - x_b^*}{x_a - x_a^*}\right) \bigg/ \ln\left(\frac{x_b - x_a}{x_b^* - x_a^*}\right).$$

So,

$$N = \ln\left(\frac{4.5 \times 10^{-4} - 1.2 \times 10^{-3}}{3.9 \times 10^{-2} - 1.0 \times 10^{-1}}\right) \bigg/ \ln\left(\frac{4.5 \times 10^{-4} - 1.0 \times 10^{-1}}{1.2 \times 10^{-3} - 1.0 \times 10^{-1}}\right) = 4.6.$$

Rounding up to the next integer $N = 5$. Including Stage 1, the number of equilibrium-limited stages is six.

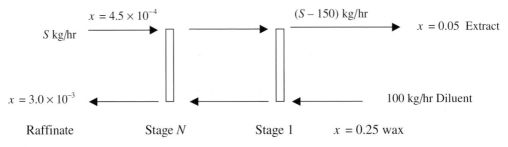

Figure 5.29 Schematic solvent flow diagram, Example 5.7.

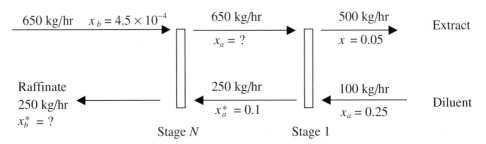

Figure 5.30 Redrawn schematic for Example 5.7.

5.12 Remember

- Extraction is a UNIT OPERATION. Regardless of what chemicals are being separated, the basic design principles for extraction are always similar.
- The assumption that stages in an extraction column are in equilibrium allows calculations of concentrations and temperatures without detailed knowledge of flow patterns and heat and mass transfer rates. This assumption is a major simplification.
- Extraction requires different solubilities of the solute in the two liquid phases. It also requires that the two phases have different densities so that they may also be separated.

- The spent solvent is often recovered (usually by distillation) and reused in the extraction process.
- Extraction can be done in a single stage, countercurrent cascades, and cross-flow cascades. Extraction columns may have distinct stages or packing, and many have highly specialized pieces of equipment to facilitate mixing.
- Accounting for efficiencies in extraction columns is much more difficult than for distillation columns, and is not considered here.
- All extraction systems are partially miscible to some extent, but when partial miscibility is very low, the system may be treated as completely immiscible and McCabe–Thiele analysis is appropriate.
- Leaching is a solid–liquid extraction and is sometimes treated similarly to partially miscible liquid–liquid extraction. The general term "extraction" is usually limited to the liquid–liquid type.
- Strengths
  - Can be done at low temperatures to protect unstable molecules
  - Can be energy efficient
  - Can separate components with azeotropes or low relative volatilities
- Weaknesses
  - Some equipment can be very complicated and/or expensive
  - Added complexities of a mass-separating agent (solvent): recovery, recycle, storage, etc.
  - More difficult to model and scale-up vapor/liquid than distillation.

5.13   Questions

5.1 In Figure 5.20, V_0 and O_0 are not on the solubility envelope while O_1, O_2, O_3, V_1, V_2, and V_3 are. Why?

5.2 In Figure 5.20, which stage would be best for addition of pure solvent? Why?

5.3 Why isn't a cocurrent cascade used in solvent extraction?

5.4 Acetic-acid extraction examples in this chapter used both isopropyl ether and 1-butanol. Which is a better solvent? Why?

5.14   Problems

5.1 Using the lever-arm rule, derive equations for O/F and V/F.

5.2 Show that the lever-arm rule is valid for one inlet and two outlet streams for a contact stage.

5.3 For acetic-acid extraction using isopropyl ether, redo the calculation using a composition of 95 wt% isopropyl ether and 5 wt% water. This composition is easier to obtain in the distillation step.

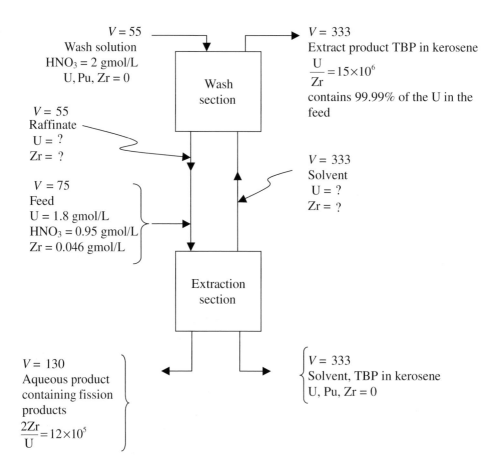

$$\text{Distribution coefficient } (K) = \frac{\text{conc. in organic phase}}{\text{conc. in aqueous phase}}$$

	HNO$_3$, gmol/L	
	2.0	1.39
K_U	12	11
K_{Pu}	8	5
K_{Zr}	0.06	0.03

Figure 5.31 Diagram and data relating to Problem 5.7.

5.4 For Example 5.3, illustrating minimum solvent flowrate, show that increasing the solvent flowrate will decrease the required number of equilibrium stages.

5.5 For the same example, show that the maximum concentration of A in O_N will occur at the minimum flowrate of solvent.

5.6 In a copper-wire plant, sodium-hydroxide solution (drag-out from a neutralization bath) is removed from wire coils by countercurrent rinsing with water. At present, the

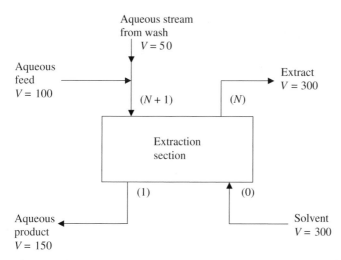

Aqueous stream
from wash
$V = 50$

Aqueous
feed
$V = 100$

Extract
$V = 300$

$(N + 1)$ (N)

Extraction
section

(1) (0)

Aqueous
product
$V = 150$

Solvent
$V = 300$

Figure 5.32 Plutonium-recovery extraction process for Problem 5.8.

drag-out solution from the neutralization bath contains 2.78×10^{-3} gmol NaOH/L and the drag-out with the product wire is at pH 9.0. Two countercurrent rinse stages are used. Fresh water, pH 7.0, is fed to the rinse operation. The wash-water-to-drag-out ratio is 16.2 (the drag-out and wash-water flowrates may be assumed constant). A process change will double the NaOH concentration in the neutralization bath and will produce drag-out at the same flowrate as before, but containing 5.55×10^{-3} gmol NaOH/L. If the same wash-to-drag-out ratio is used as before, how many wash stages would be needed to achieve:

(a) the same percentage reduction of the NaOH concentration in the drag-out as at present;

(b) the same NaOH concentration in the drag-out with the product wire as at present?

5.7 An extraction process for separating actinide elements (principally uranium, U, and plutonium, Pu) from fission products in an aqueous solution of spent fuel rods is illustrated in Figure 5.31. The extraction solvent is 30% tributyl phosphate (TBP) in kerosene. The most extractable of the fission products are zirconium, niobium and ruthenium. Zirconium, Zr, is used herein to represent the fission products. Determine the number of stages required in the wash section and in the extraction section. Determine the percentage of the Pu in the feed which is recovered in the extract product. V denotes the relative volumetric flowrate.

5.8 Consider the plutonium-recovery extraction process shown in Figure 5.32. Determine the additional number of stages required for each ten-fold reduction in plutonium concentration, C, in stream 1 (for conditions of high plutonium recovery, i.e., $C(1) \ll C(N + 1)$). The plutonium distribution coefficient, K_{Pu}, is 5 (ratio of concentration in organic phase to concentration in aqueous phase). V denotes the relative volumetric flowrate.

6

Absorption and stripping

Prevention is better than cure.

– ERASMUS

6.1 Objectives

1 Calculate a mass and a mole ratio and be able to convert from mass (or mole) fractions to mass (or mole) ratios.
2 Calculate a column diameter based upon flooding velocity.
3 Use the McCabe–Thiele graphical method to determine number of equilibrium stages and outlet concentrations for absorption and stripping processes.
4 Calculate the number of equilibrium stages and/or outlet concentrations using analytical methods for dilute concentrations.
5 Use the HTU–NTU (height of transfer unit–number of transfer units) method to determine the height of packing in:
   – concentrated solutions in absorption and stripping columns
   – dilute solutions in absorption and stripping columns.

6.2 Background

Absorption is the process by which one or more components of a gas phase are removed by being solubilized (sorbed) into a liquid. The liquid phase is the mass-separating agent (MSA); it allows separation by absorption to occur. Stripping, or desorption, is the opposite of absorption. A reasonably volatile contaminant is moved from a liquid to a gas phase. In either case the contaminant to be removed is called the solute and the MSA which does the removal is called the solvent. During absorption, a low relative volatility, K, value as defined in Chapter 3 is beneficial. Stripping works best at conditions that yield a high K value.

Absorption can be either physical or chemical. Physical absorption takes advantage of solubility differences of the gaseous component to be removed between the gas and the liquid phases. Chemical absorption (which is not covered here) requires the gas to be removed to react with a compound in the liquid solvent and to remain in solution. Chemical absorption involves irreversible and reversible reactions. Reversible reactions make it possible to recover and recycle the liquid solvent, while irreversible processes allow for only a one-time solvent use.

The most common method of treating wastewaters to reduce the level of organic contaminants is steam stripping, particularly when the contaminant's boiling point is lower than the boiling point of water (e.g., methylene chloride, acetone, methanol, benzene, and toluene) [1].

One environmental example of absorption is the removal of ammonia gas from an air stream with water as the mass-separating agent. With the proper choice of a liquid solvent MSA, acid gases and other contaminants can also be removed from air streams using absorption processes. Air-stripping towers used to remove volatile organic compounds (VOCs) from water are a common application of stripping. Again, with the right gas phase MSA, many volatile contaminants can be removed from liquid streams.

The primary advantage of absorption and stripping processes is that they usually do not require reboilers or condensers. One exception is that heat input to a stripping column can often make a non-volatile contaminant adequately volatile for removal by the solvent. Another advantage is that the absorption solvent can often be recycled; usually through the use of a stripping column to recover the solvent. A weakness of these technologies is that chemical absorption typically has fairly low Murphree efficiencies for staged columns.

To optimally remove a solute from a feed stream with a solvent in either absorption or stripping, it is best to have the highest possible contact between the two phases for an extended period of time. This can be achieved in one of two column designs: equilibrium stage or packed. The gas flows from the bottom of the column to the top and the liquid falls countercurrent from the top down in both designs. Spray columns, which are a somewhat simplified form of packed towers, are also occasionally used. The mathematical characterization of the process is different for staged and packed column design; staged columns are analyzed with equilibrium as the controlling mass transfer mechanism while packed columns are analyzed in terms of resistance to mass transfer between the phases. Both analyses will be presented in this chapter.

In general, continuously packed columns are basically identical whether they are used for absorption or stripping applications. Kohl provides a helpful guide for choosing the best type of equipment for various stripping and absorption systems (Table 6.1) [2].

Table 6.1 *Selection guide for absorbers and strippers [2].*

	Tray columns		Packed columns
	Perforated	Bubble cap	(Random)
Low liquid flowrate	D	A	C
Medium liquid flowrate	A	C	B
High liquid flowrate	B	C	A
Difficult separation (many stages)	A	B	A
Easy separation (one stage)	C	C	B
Foaming system	B	C	A
Corrosive fluids	B	C	A
Solids present	B	D	C
Low pressure difference	C	D	B
High turndown ratio	C	A	B
Versatility (can be changed)	C	C	A
Multiple feed and drawoff points	B	A	C

A = best choice, D = worst choice

6.3 Column diameter

Regardless of whether equilibrium or mass transfer analysis is used to design an absorption column, the column diameter is specified according to some limiting flow conditions for the gas phase. There is a maximum gas flowrate that corresponds to any liquid flowrate. This maximum gas velocity is called the flooding velocity and above it the phenomenon called flooding begins to occur. When the gas velocity exceeds the flooding velocity, the liquid is held up. In other words, it ceases to fall. It is literally supported on top of the gas stream such that greater liquid pressure is required to force the liquid through the column. Eventually, if the gas velocity becomes too high, liquid accumulates at the top of the column and is blown out with the exiting gas stream.

The gas velocity must, therefore, be less than the flooding velocity. However, if it is much less, the mass transfer between the two phases will be less efficient and, hence, taller columns will be required to perform a given separation. The advantage of low gas velocities is lower liquid pressure losses and, thus, lower pumping cost to operate the column. A balance in gas velocity is needed between the liquid pumping power costs and the fixed cost and practicality of excessively tall columns. Typically, a column is designed to operate at one-half the flooding velocity.

Because the gas throughput (mass or volumetric flow) in an absorption process is typically a set design parameter, a variable other than gas flow must be adjusted to maintain an acceptable velocity. The column diameter is the chosen parameter. According to

$$Q = v^* A,$$

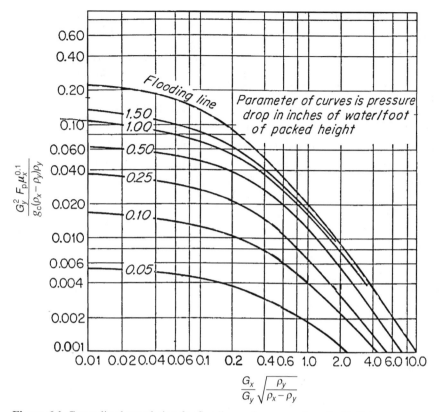

Figure 6.1 Generalized correlation for flooding and pressure drop. L. McCabe, *Unit Operations of Chemical Engineering*. Reproduced with permission of McGraw-Hill companies.

where Q is volumetric flowrate (ft^3/min), v is velocity (ft/min), and A is area (ft^2), increasing the diameter for a constant volumetric flowrate results in a lower velocity. Figure 6.1 is a logarithmic plot of

$$\frac{G_y^2 F_p \mu_x^{0.1}}{g_c(\rho_x - \rho_y)\rho_y} \quad \text{vs} \quad \frac{G_x}{G_y}\sqrt{\frac{\rho_y}{\rho_x - \rho_y}},$$

where G_x = mass velocity of liquid, lb/ft$^2 \cdot$ s
$\quad G_y$ = mass velocity of gas, lb/ft$^2 \cdot$ s
$\quad F_p$ = packing factor, ft^{-1}
$\quad \rho_x$ = density of liquid, lb/ft^3
$\quad \rho_y$ = density of gas, lb/ft^3
$\quad \mu_x$ = viscosity of liquid, cP
$\quad g_c$ = Newton's-Law proportionality factor, 32.174 ft $\cdot$ lb/lb$_f \cdot$ s^2.

For a given gas with solute and liquid solvent pair, the G_x/G_y or L/G ratio can be set using a comparison of the equilibrium line to the operating line. This concept will be discussed later in the chapter. These two ratios are equal because the area term cancels between the numerator and denominator of the first ratio, reducing it to the second. Then, because density information for both streams can be obtained for a column operating at constant temperature and pressure, the x-axis of the figure can be calculated. The ideal gas law can be used for the gas-phase density calculation and the liquid-phase density can be found in many tables as a function of temperature (assuming an incompressible fluid). One can then read off the figure the value of the y-axis that corresponds to flooding. Again, every quantity here except for the gas mass velocity is a constant at a given temperature and pressure. A list of packing factors is supplied in Table 6.2, p. 170. The gas mass velocity can be calculated and a simple ratio of the gas mass flowrate to the gas mass velocity will give the required cross-sectional area of the column:

$$\text{area} = \text{mass flow/mass velocity.} \tag{6.1}$$

More details on selection and scale-up of columns are available [3].

Example 6.1: flooding velocity column diameter

Problem:

An absorption column is to be built to separate ammonia from air using ammonia-free water as a solvent. A tower packed with 1-inch ceramic Rasching rings needs to treat 25,000 cubic feet per hour of entering gas with 2% ammonia by volume. The column temperature is 68 °F and the pressure is 1 atm. The ratio of gas flowrate to liquid flowrate is:

1.0 lb gas/1.0 lb liquid.

If the gas velocity is one-half the flooding velocity, determine the necessary column diameter.

Solution:

For H_2O,

$$\rho_x = 62.4 \text{ lb/ft}^3$$

$$\mu_x = 1.0 \text{ cP.}$$

For air, molecular weight

$$M = 0.98(29) + 0.02(17) = 28.76 \text{ lb/lbmol.}$$

Then,

$$\frac{G_x}{G_y}\left(\frac{\rho_y}{\rho_x - \rho_y}\right)^{1/2} = 1.0\left(\frac{0.75}{62.4 - 0.075}\right)^{1/2} = 0.0346.$$

For flooding, from Figure 6.1,

$$\frac{G_{y2}F_p\mu_x^{-1}}{g_c(\rho_x - \rho_y)\rho_y} = 0.19; \qquad F_p = 155 \text{ ft}^{-1}$$

$$G_y^2 = \frac{0.19}{} \left| \frac{32.2 \text{ ft} \cdot \text{lb/lbmol}}{\text{lb}_f \cdot \text{s}^2} \right| \frac{62.3 \text{ lb}}{\text{ft}^3} \left| \frac{0.075 \text{ lb}}{\text{ft}^3} \right| \frac{}{155} \left| \frac{}{0.1} \right.$$

$$G_y = \frac{0.43 \text{ lb}}{\text{ft}^2 \cdot \text{s}}.$$

Then:

$$\text{Area} = \frac{\text{mass flowrate (lb/s)}}{\text{mass velocity (lb/ft}^2 \cdot \text{s)}}.$$

Mass flowrate:

$$\frac{25{,}000 \text{ ft}^3}{\text{hr}} \left| \frac{0.075 \text{ lb}}{\text{ft}^3} \right| \frac{\text{hr}}{3600 \text{ s}} = 0.518 \frac{\text{lb}}{\text{s}}.$$

We want one-half the flooding velocity, so:

$$\text{Mass velocity} = \frac{0.43 \text{ lb}}{\text{ft}^2 \cdot \text{s}} \left| \frac{0.5}{} \right. = 0.215 \frac{\text{lb}}{\text{ft}^2 \cdot \text{s}},$$

$$\text{Area} = \frac{0.518 \text{ lb}}{\text{s}} \left| \frac{\text{ft}^2 \cdot \text{s}}{0.215 \text{ lb}} \right. = 2.42 \text{ ft}^2,$$

$$\text{Diameter:} \quad 2.42 \text{ ft}^2 = \frac{\pi d^2}{4};$$

$$d = 1.76 \text{ ft}.$$

6.4 McCabe–Thiele analysis: absorption

6.4.1 Operating lines (mass balances)

Figure 6.2 shows a basic flow diagram for either an absorption or a stripping column. The process differs from distillation in that one stream enters with a contaminant and exits "clean" while a second enters "clean" and exits with the contaminant. Hence, there are two feed streams, the solute carrier and the mass-separating agent, which enter the column at opposite ends. This creates a flow pattern similar to distillation in which a gas phase flows countercurrent to a liquid phase. An absorption column is equivalent to the rectifying section of a distillation column.

Using the assumptions that the liquid is non-volatile and the carrier gas is insoluble, the mass balances (total and solute) are:

$$L_N = L_j = L_0 = L = \text{constant} \qquad \text{(liquid balance)}, \tag{6.2}$$

$$G_{N+1} = G_j = G_1 = G = \text{constant} \quad \text{(carrier gas mass balance)}. \tag{6.3}$$

It follows naturally that we can also say:

$$\frac{L}{G} = \frac{\text{moles non-volatile liquid/hr}}{\text{moles insoluble carrier gas/hr}} = \text{constant}. \tag{6.4}$$

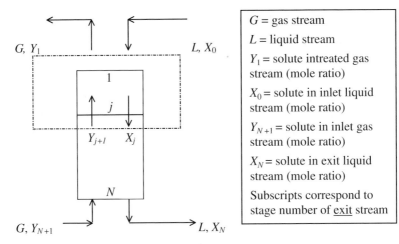

G = gas stream

L = liquid stream

Y_1 = solute intreated gas stream (mole ratio)

X_0 = solute in inlet liquid stream (mole ratio)

Y_{N+1} = solute in inlet gas stream (mole ratio)

X_N = solute in exit liquid stream (mole ratio)

Subscripts correspond to stage number of <u>exit</u> stream

Figure 6.2 Absorption column schematic.

We cannot, however, say that overall flowrates of gas and liquids are constant (except for very dilute solutions) because a significant amount of solute may be absorbed. This would increase the total flowrate of liquid, while reducing that of the gas. The compositions of the solute must therefore be expressed in mole ratios so that the basis (denominator) is constant. Mole ratios are related to mole fractions by the equations:

$$Y = \frac{y}{1-y} = \frac{\text{moles solute}}{\text{moles insoluble carrier gas}}$$

and

$$X = \frac{x}{1-x} = \frac{\text{moles solute}}{\text{moles pure liquid}}.$$

So that the solute balance is:

$$Y_{j+1}G + X_0L = X_jL + Y_1G. \tag{6.5}$$

Solving for Y_{j+1} gives:

$$Y_{j+1} = \frac{L}{G}X_j + \left(Y_1 - \frac{L}{G}X_0\right), \tag{6.6}$$

the operating line for absorption. As in distillation analysis, an operating line can be plotted if L, G, and a single point are known. The operating line is a straight line with slope L/G on a Y vs X diagram. The equilibrium line can be curved. For absorption, the Y intercept is greater than zero so the operating line is *above* the equilibrium line for this type of plot.

Note that two additional assumptions are necessary to neglect the energy balances for absorption processes. They are that the heat of absorption is negligible and that the operation is isothermal.

6.4.2 Procedure for solving absorption problems graphically

The graphical solution to absorption is identical to that for the rectifying section of a distillation column, except that the operating line is now above the equilibrium line. This is because the solute is now being transferred from the gas to the liquid instead of from the liquid to the gas. The "stepping off" stages procedure is as follows.

1 Plot Y vs X equilibrium data (make sure to convert from mole fractions to mole ratios!).
2 The values X_0, Y_{N+1} (inlet solute concentrations or similar values), and L, G (flowrates) are usually given. One point on the operating line is represented by the passing streams: (X_0, Y_1), where Y_1 is the target exit value. Using this point and a slope of L/G, plot the operating line.
3 Step off stages as usual, starting at Stage 1. Start at operating line (X_0, Y_1) and move horizontally to equilibrium line (X_1, Y_1). Move vertically to operating line (X_1, Y_2). Repeat this procedure until you reach the opposite end of column on operating line.

6.4.3 Limiting liquid/gas ratio

Analogous to the minimum reflux ratio in distillation, there exists a minimum L/G ratio in absorption. Figure 6.3 compares the operating line to the equilibrium line for absorption. Because the slope of the operating line is L/G, reducing the liquid mass-separating agent flowrate decreases the slope of the line. Compare lines aA and ab for a separation in which the gas enters with Y_b concentration of solute which must be removed to Y_a and the liquid

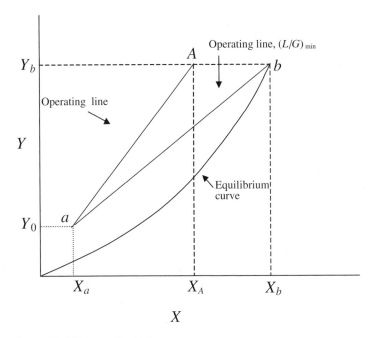

Figure 6.3 Minimum liquid flow.

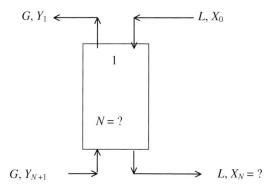

Figure 6.4 Schematic for absorption column, Example 6.2.

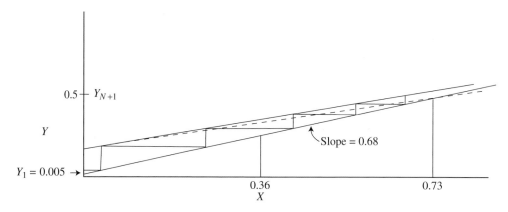

Figure 6.5 Graphical determination of equilibrium stages, Example 6.2.

solvent enters with X_a solute concentration. The maximum exiting liquid concentration and the corresponding minimum liquid flow rate are given by line *ab*.

Because *ab* intersects the equilibrium line, no further separation can occur. In other words, the equilibrium concentration of the solute in the solvent cannot be exceeded. This is the minimum liquid flowrate. Just as with the minimum reflux ratio, this situation is hypothetical because an infinite number of equilibrium stages (or an infinitely tall column) would be required.

Despite its theoretical nature, it is still useful to know the minimum liquid flowrate because it corresponds to the maximum concentration of the solute in the mass separating agent (largest value of X). This becomes important when the solvent MSA is recovered downstream for recycle back into the system. When the solute is highly concentrated in the MSA, there is less total material that must be processed in the solvent recovery system. When the solute is dilute in the MSA, more material must be processed in the solvent recovery system. However, fewer theoretical stages are required in the absorption column. A good design will balance these two factors. Also, a typical design will specify the actual L/G as a multiple of the minimum value.

Example 6.2: absorption column (Figure 6.4)

Problem:

A vent gas stream in a chemical plant is 15 wt% *A*; the rest is air. The local pollution authorities feel that *A* is a priority pollutant and require a maximum exit concentration of 0.5 wt%. It is decided to build an absorption tower using water as the absorbent. The inlet water is pure and at 25 °C. At 25 °C, the laboratory has found that the equilibrium data can be approximated by $y = 0.5x$ (where y and x are mass fractions of *A* in vapor and liquid, respectively).

(a) Find the minimum ratio of water to air $(L/G)_{min}$ (on an *A*-free basis).

(b) With an $L/G = 1.5(L/G)_{min}$, find the total number of equilibrium stages.

Solution:

The first step is to plot the equilibrium curve using the given relationship (see Figure 6.5). Since the known variables are given in mass fractions, the data will be graphed as mass ratios instead of mole ratios. Rearranging the equilibrium equation in terms of mass ratios

$$y = 0.5x$$

$$Y = \frac{y}{1-y} \Rightarrow y = \frac{Y}{1+Y}$$

$$X = \frac{x}{1-x} \Rightarrow x = \frac{X}{1+X}.$$

Substituting:

$$\frac{Y}{1+Y} = 0.5\frac{X}{1+X}.$$

Calculate exit concentrations as mass ratios:

$$y_{N+1} = 0.05 \qquad Y_{N+1} = \frac{0.05}{1-0.05} = 0.5$$

$$y_1 = 0.005 \qquad Y_1 = \frac{0.005}{1-0.005} = 0.005.$$

Over this range of *Y*, the equilibrium line will be linear $Y = 0.5X$.

The next step is to find $(L/G)_{min}$. This is the case where the column would have infinite stages, and corresponds to a pinch point. To draw the operating line with minimum slope, connect (X_0, Y_1) and the point where the equilibrium line crosses the value Y_{N+1}. This second point is easily found from the equilibrium data and the two points are used to find the slope:

$$\left(\frac{L}{G}\right)_{min} = \frac{Y_{N+1} - Y_1}{X_N - X_0} = \frac{0.05 - 0.005}{0.1 - 0} = \frac{0.045}{0.1} = 0.45$$

$$1.5\left(\frac{L}{G}\right)_{min} = 1.5(0.45) = 0.68.$$

The actual operating line was plotted with a slope of $1.5(L/G)_{min} = 0.68$, also starting from the point (X_0, Y_{N+1}). The result is four stages. Note that four stages provides a better separation than required. This means that the actual Y_1 value will be below the target value.

6.5   McCabe–Thiele analysis: stripping

Stripping is very similar in concept to absorption, but mass transfer occurs in the opposite direction: it is the transfer of a component from a liquid stream into a gas stream mass-separating agent. The mass balances and the operating line are derived in a similar fashion to those for absorption. Referring to Figure 6.6, the operating line is

$$Y_n = \left(\frac{L}{G}\right) X_{n+1} + \left(Y_0 - \frac{L}{G} X_1\right). \tag{6.7}$$

The Y intercept is less than zero so the operating line is below the equilibrium line.

One useful limit for a stripping column is the maximum L/G ratio, which corresponds to the minimum stripping gas flowrate (minimum G/L ratio) required for a desired separation. The maximum L/G is the slope of the line which begins at the point (Y_0, X_1) and intersects the equilibrium line. As shown in Figure 6.7, the $(L/G)_{max}$ occurs as one rotates the operating line clockwise around (X_1, Y_0) until it intersects the equilibrium line. This can occur at a tangent pinch point and not necessarily at the end of the column (X_N, Y_N).

Example 6.3: stripping column

Problem:
A volatile organic carbon (VOC) in a water stream is to be stripped out using an air stream in a countercurrent staged stripper. Inlet air is pure, and flowrate is $G = 500$ lb/hr. Inlet liquid stream has a mass ratio of $X = 0.10$ and a flowrate of pure water of 500 lb/hr. The desired outlet mass ratio is $X = 0.005$. Assume that water is non-volatile and air is insoluble. Find the number of equilibrium stages and the outlet gas mass ratio. Equilibrium data can be represented by $Y = 1.5 X$.

Solution:
Figure 6.8 is a schematic diagram of the column. The equilibrium data are simple to plot (Figure 6.9) because everything is already given in terms of mass ratios. One point on the operating line is (X_1, Y_0), and the slope is 1 since the inert liquid and gas flowrates are equal. The separation will require five equilibrium stages, and the outlet gas mass ratio (Y_N) is 0.095.

6.5.1   Analytical methods

When the solute concentration in both the gas and liquid phase is very dilute, the total flowrates can be considered constant. The mole or mass ratios reduce to the corresponding

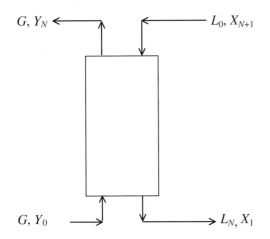

Figure 6.6 Stripping column schematic.

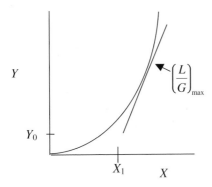

Figure 6.7 Minimum gas flowrate.

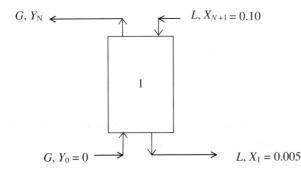

Figure 6.8 Schematic for stripping column, Example 6.3.

mole or mass fractions. For isothermal operation and no solute in the entering liquid phase ($x_0 = 0$), the Kremser equation, Equation (3.49), becomes

$$\frac{y_{N+1} - y_1}{y_{N+1} - 0} = \frac{(L/mV) - (L/mV)^{N+1}}{1 - (L/mV)^{N+1}}.$$

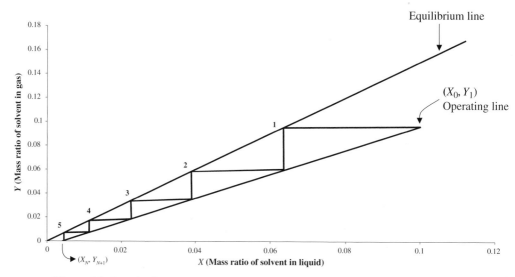

Figure 6.9 X–Y plot for stripping column, Example 6.3.

The term $L/mV = A$ is called the absorption factor. It is often written as L/KV where $y = Kx$. It would be useful to determine the ratio

$$\frac{y_1}{y_{N+1}} = \text{fraction of solute in entering gas } not \text{ adsorbed.}$$

The above equation can be rearranged to obtain:

$$\frac{y_1}{y_{N+1}} = \frac{A - 1}{A^{N+1} - 1}. \tag{6.8}$$

For absorption, a large value of A is better. Since m (or K) is in the denominator, a small value of m (or K) means a small mole fraction in the gas phase (y) relative to the value of x. Solute favors the liquid phase.

For stripping, an analogous equation can be derived:

$$\frac{x_1}{x_{N+1}} = \frac{S - 1}{S^{N+1} - 1} = \text{fraction of entering solute in liquid } not \text{ stripped,} \tag{6.9}$$

where $S = \dfrac{mV}{L} = \dfrac{KV}{L} = \dfrac{1}{A}$.

In this case, a large value of S is better. This translates to a large value of m (or K). Can you see why?

Example 6.4

Problem:

Referring to Example 6.3 for a stripping column, $L = 500\,\text{lb/hr}$, $V = 500\,\text{lb/hr}$, $m = 1.5$ and $x_{N+1} \cong 0.1$. This value of x_{N+1} is somewhat higher than what is considered dilute but it is still reasonable to use the analytical approach as a first estimate. For $N = 5$, what is x_1?

Solution:

$$S = \frac{mV}{L} = 1.5$$

$$\frac{x_1}{0.1} = \frac{1.5 - 1}{(1.5)^6 - 1}$$

$$x_1 = 4.8 \times 10^{-3}.$$

This result is very close to the value of $x_1 = 5 \times 10^{-3}$.

6.6   Packed columns

Packed towers for absorption and stripping are often used instead of columns with discreet trays. The column configuration is the same as a staged column with the liquid entering the top of the tower and flowing countercurrent to the gas stream, which enters the bottom of the column. A liquid distributor spreads the liquid over the entire cross-section of packing. Ideally, the liquid should form a thin layer over the packing surface. Most packing is shaped like saddles and rings, although some structured packings are available. Typically, the packing is dumped into the column.

The choice of packing is critical to the successful operation of the absorption or stripping process and must meet several criteria. Figure 6.10 shows some common types of packing material. The saddles and rings can be made of plastic, metal, or ceramic. For optimal liquid–gas contact the packing should be a shape with a high specific surface (surface area to volume ratio). Thus, the packing provides a large area of contact between the two phases. It must be chemically inert to all process fluids under the range of operating conditions. The packing should be strong, lightweight and low cost. It must allow adequate passage of both liquid and gas without liquid hold-up or excessive pressure drop. Table 6.2 gives some physical characteristics of commercially available packing.

6.6.1   Mass transfer principles

Figure 6.11 shows the process of mass transfer as a gas solute exits the gas phase and enters the liquid MSA phase. There are resistances to mass transfer describing movement of the

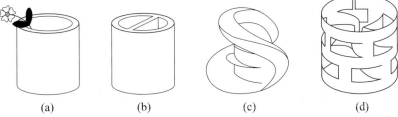

(a)   (b)   (c)   (d)

Figure 6.10 Typical tower packings: (a) Raschig ring, (b) Lessing ring, (c) Berl saddle, and (d) Pall ring [5]. Reprinted by Permission of John Wiley and Sons, Inc. Copyright 1980.

Table 6.2 *Physical characteristics of dry commercial packings (after [5]). Reprinted by permission of John Wiley and Sons, Inc. Copyright © 1980.*

Packing in (mm)	Percent voids (ϵ)	Specific surface (a_v) m²/m³	Dumped weight kg/m³ (lb/ft³)	Packing factor (F)
Ceramic Raschig rings				
¼ (6.35)	73	787	737 (46)	1600
½ (12.7)	63	364	865 (54)	580
1 (25.4)	73	190	641 (40)	155
2 (50.8)	74	92	609 (38)	65
Metal Raschig (rings (1/16 inch wall)				
½ (12.7)	73	387	2114 (132)	410
1 (25.4)	85	187	1137 (71)	137
2 (50.8)	92	103.0	59 (37)	57
Berl saddles				
¼ (6.35)	60	899	897 (56)	900
½ (12.7)	63	466	865 (54)	240
1 (25.4)	69	249	721 (45)	110
2 (50.8)	72	105	641 (40)	45
Pall rings				
1 (25.4)	93.4	217.5	529 (33)	48
2 (50.8)	94.0	120	440.5 (27.5)	20

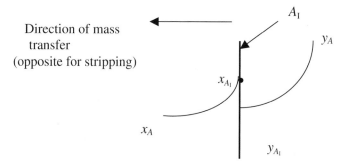

Figure 6.11 Mass transfer at interface between liquid (*left*) and vapor (*right*) phases.

solute both through the gas phase to the liquid/gas interface and across the interface into the liquid phase. When one can assume equimolar counter-transfer of the more volatile and less volatile components (a good assumption for, <u>dilute</u> absorbers and strippers), the rate of mass transfer is:

Rate $= A_1 k_y (y_A - y_{A_1})$ for the vapor phase, and

Rate $= A_1 k_x (x_{A_1} - x_A)$ for the liquid phase.

A_I is the interfacial area between the liquid and vapor phases, x_{A_I} and y_{A_I} are the interfacial mole fractions, x_A and y_A are the bulk-phase mole fractions, and k_y are the individual mass transfer coefficients for the liquid and vapor phases, respectively.

Since the interfacial area, A_I, is difficult to determine accurately, these rate equations may be rewritten as:

$$\frac{\text{Rate}}{\text{Volume}} = k_y a(y_A - y_{A_I}) = k_x a(x_{A_I} - x_A), \tag{6.10}$$

where a is the specific surface of the column packing. It is also difficult to determine a accurately, so values are correlated into lumped term mass transfer correlations for $k_y a$ and $k_x a$. It is also possible to write the equation without the interfacial mole fractions by using overall coefficients K_y and K_x:

$$\frac{\text{Rate}}{\text{Volume}} = K_y a(y_A - y_A^*) = K_x a(x_A^* - x_A). \tag{6.11}$$

Here, x_A^* is the liquid mole fraction that would be in equilibrium with y_A, and y_A^* is the vapor mole fraction that would be in equilibrium with x_A.

Since the mass transfer occurs in series, we can derive an overall mass transfer coefficient in terms of mass transfer in the gas phase:

$$\frac{1}{K_y a} = \frac{m}{k_x a} + \frac{1}{k_y a}. \tag{6.12}$$

Similarly, in terms of mass transfer in the liquid phase:

$$\frac{1}{K_x a} = \frac{1}{k_x a} + \frac{1}{m k_y a}. \tag{6.13}$$

This form should look familiar; it is the sum-of-resistances model from Chapter 3. If m is small, $K_y \approx k_y$ and the gas-phase resistance is the dominant. If m is large, $K_x \approx k_x$ and the liquid-phase resistance controls. Remember what a small or large value of m implies for the solubility of a solute in the liquid phase relative to the gas-phase concentration.

Another approach to understanding the controlling resistance is to rearrange the equation for the rate/volume in each phase:

$$-\frac{k_x a}{k_y a} = \frac{y_A - y_{A_I}}{x_A - x_{A_I}}. \tag{6.14}$$

This can be rewritten as

$$-\frac{1/k_y a}{1/k_x a} = \frac{y_A - y_{A_I}}{x_A - x_{A_I}} = -\frac{\text{gas-phase mass transfer resistance}}{\text{liquid-phase mass transfer resistance}}. \tag{6.15}$$

If gas-phase resistance is negligible, $y_A \approx y_{A_I}$. Similarly, $x_A \approx x_{A_I}$ if the liquid-phase resistance can be neglected.

6.6.2   General HTU–NTU method

The height of the packing in a particular column for a given separation can be calculated with the HTU–NTU method. This method is based on mass transfer between the vapor and liquid phases.

Assuming constant molar overflow, L/V is constant and the assumption of equimolar counterdiffusion is valid, so that the flux of one component across the vapor–liquid interface is equal and opposite to the flux of the other component ($N_A = -N_B$).

For a differential height dz in a packed column, the mass transfer rate is:

$$N_A a A_c dz = k_y a (y_A - y_{A_I}) A_c dz, \tag{6.16}$$

where N_A is the flux of solute A, and A_c is the column cross-sectional area. The mass transfer rate in terms of changes of solute concentration in the liquid and vapor phases is:

$$N_A a A_c dz = -V dy_A = L dx_A \quad (L \text{ and } V \text{ are constant}). \tag{6.17}$$

The last two equations may be combined to give:

$$dz = \frac{V}{k_y a A_c (y_{A_I} - y_A)} dy_A. \tag{6.18}$$

Integrating from $z = 0$ to $z = \ell$ (where ℓ is the total height of packing in a stripping/enriching section):

$$\ell = \frac{V}{k_y a A_c} \int_{y_{A_{in}}}^{y_{A_{out}}} \frac{dy_A}{y_{A_I} - y_A}. \tag{6.19}$$

This equation is often written as:

$$\ell = H_G N_G, \tag{6.20}$$

where

$$H_G = \frac{V}{k_y a A_c} = \text{height of a gas-phase transfer unit (HTU)}$$

and

$$N_G = \int_{y_{A_{in}}}^{y_{A_{out}}} \frac{dy_A}{y_{A_I} - y_A} = \text{number of gas-phase transfer units (NTUs)}.$$

The number of transfer units is a measure of the difficulty of a separation. The higher the purity of the product, the larger the number of transfer units. This is analogous to the need for more equilibrium stages in distillation to obtain a higher-purity product. The height of a transfer unit is a measure of the efficiency of a particular column packing and configuration to perform the separation. The smaller the value of HTU, then the higher the effectiveness of the separation for a given transfer unit.

A similar model is obtained from the liquid phase:

$$\ell = \frac{L}{k_x a A_c} \int_{x_{A_{out}}}^{x_{A_{in}}} \frac{dx_A}{x_A - x_{A_I}} = H_L N_L. \tag{6.21}$$

To find N_G and N_L, we can return to:

$$-\frac{k_x a}{k_y a} = \frac{y_A - y_{A_I}}{x_A - x_{A_I}}.$$

This equation is the slope of a line from (y_A, x_A) to (y_{A_I}, x_{A_I}), the bulk and interfacial mole fractions, respectively. The interfacial mole fractions are in equilibrium, and so the point (y_{A_I}, x_{A_I}) is on the equilibrium line. The point (y_A, x_A) is on the operating line.

From any point on the operating line (y_A, x_A), a line with slope $-k_x a/k_y a$ will intersect the equilibrium line at the corresponding (y_{A_I}, x_{A_I}). This can be done for a number of points, and a plot of $1/(y_{A_I} - y_A)$ vs y_A can be generated. The area under the curve between $y_{A_{in}}$ and $y_{A_{out}}$ will be N_G for the section. N_L can be found similarly by plotting $1/(x_A - x_{A_I})$.

It would be useful to eliminate the need to determine the interfacial compositions and just use the bulk concentrations in each phase. This can be done by using an overall mass transfer coefficient along with the appropriate driving force. The same derivation procedure results in

$$\ell = \frac{V}{K_y a A_c} \int_{y_{A_{out}}}^{y_{A_{in}}} \frac{dy}{y_A - y_A^*} = H_{OG} N_{OG}. \tag{6.22}$$

A complete listing of these terms is shown in Table 6.3 (see p. 176).

If we assume linear phase equilibrium $(y = mx)$ and a linear operating line, an analytical expression can be obtained for N_{OG}:

$$N_{OG} = \int_{y_{A_{out}}}^{y_{A_{in}}} \frac{dy}{(1 - mV/L)y + y_{A_{out}}(mV/L) - mx_{A_{in}}}. \tag{6.23}$$

Integrating and using the definition of the absorption factor A:

$$N_{OG} = \frac{\ln \left\{ \left[\frac{(A-1)}{A} \right] \left[\frac{y_{A_{in}} - mx_{A_{in}}}{y_{A_{out}} - mx_{A_{in}}} \right] + \frac{1}{A} \right\}}{\frac{(A-1)}{A}}. \tag{6.24}$$

An analogous derivation based on liquid-phase concentrations yields:

$$\ell = H_{OL} N_{OL}$$

$$H_{OL} = \frac{L}{K_x a A_c} \tag{6.25}$$

$$N_{OL} = \frac{\ln \left\{ (1 - A) \left[\frac{x_{A_{in}} - y_{A_{in}}/m}{x_{A_{out}} - y_{A_{in}}/m} \right] + A \right\}}{(1 - A)}.$$

6.6.3 Concentrated absorbers and strippers

For concentrated absorbers and strippers, the total flowrates are not constant. Solute transfer is diffusion through a stationary component ($N_B = 0$). The amount of solute transferred is sufficient to change the total flowrates *and* the driving force for mass transfer as a function of position in the column. Thus, a log mean driving force (ℓm) is the appropriate choice.

The rate of mass transfer can be defined to look like the one given previously for dilute solutions:

$$N_A a = k'_y a (y_A - y_{A_I}).$$

The mass transfer coefficient is now: $k'_y = k_y a / (1 - y_A)_{\ell m}$, where k'_y can be thought of as a mass transfer coefficient adjusted by using a mean (average) mole fraction. In this case, the mean used is the log mean to account for the variation in driving force throughout the column. Hence,

$$(1 - y_A)_{\ell m} = \frac{(1 - y_A) - (1 - y_{A_I})}{\ln \left(\dfrac{1 - y_A}{1 - y_{A_I}} \right)}. \tag{6.26}$$

Repeating the analysis for dilute solutions, but now accounting for variable total flowrates,

$$L' = L(1 - x_A)$$

$$V' = V(1 - y_A), \tag{6.27}$$

where L' and V' are the constant flowrates of inerts in each phase.

In this case the differential in concentration is

$$d(V y_A) = d \left(V' \frac{y_A}{1 - y_A} \right) = V' d \left(\frac{y_A}{1 - y_A} \right) = V' \frac{dy_A}{(1 - y_A)^2} = V \frac{dy_A}{(1 - y_A)}. \tag{6.28}$$

The resulting equation for the column height is (in terms of overall mass transfer coefficient):

$$\ell = \int_{y_{A_{\text{out}}}}^{y_{A_{\text{in}}}} \frac{V}{K'_y a A_c} \frac{dy_A}{(1 - y_A)(y_A - y_A^*)}. \tag{6.29}$$

Rearranging,

$$\ell = \int_{y_{A_{\text{out}}}}^{y_{A_{\text{in}}}} \frac{V}{K'_y a A_c (1 - y_A)_{\ell m}} \frac{(1 - y_A)_{\ell m} \, dy_A}{(1 - y_A)(y_A - y_A^*)}$$

$$= \frac{V}{K'_y a A_c (1 - y_A)_{\ell m}} \int_{y_{A_{\text{out}}}}^{y_{A_{\text{in}}}} \frac{(1 - y_A)_{\ell m} \, dy_A}{(1 - y_A)(y_A - y_A^*)} \tag{6.30}$$

and,

$$H_{OG} = \frac{V}{K_y' a A_c \, (1 - y_A)_{\ell m}}. \tag{6.31}$$

H_{OG} can be taken out of the integral (assumed constant) if average values of V and $(1 - y_A)_{\ell m}$ are used. This is normally a good assumption based on the error with correlations for mass transfer coefficients.

Table 6.3 summarizes equations for HTUs and NTUs. The choice can be dictated by the form of the mass transfer coefficient used and the phase which contributes the limiting resistance. Each form of the mass transfer coefficient has a corresponding driving force associated with it. The important point is that one uses the proper equation for both the NTU and HTU terms.

We can derive a relationship between H_{OG}, H_G, and H_L. With the assumption of dilute solutions,

$$\frac{1}{K_y a} = \frac{1}{k_y a} + \frac{m}{k_x a}$$

$$\frac{V}{k_y a A_c} = \frac{V}{k_y a A_c} + \frac{m V}{k_x a A_c}$$

$$= \frac{V}{k_y a A_c} + \frac{m V}{L} \frac{L}{k_x a A_c},$$

we obtain:

$$H_{OG} = H_G + \frac{m V}{L} H_L.$$

Correlations

Equations (6.32) and (6.33) below can be used to estimate H_G and H_L for a preliminary design [6]. More detailed correlations are available [2, 6, 7] for more accurate calculations. The advantage of the correlations below are simplicity of use and calculation which is often a benefit when doing an initial design calculation.

$$H_G = a_G W_G^b Sc_G^{0.5} / W_L^c \tag{6.32}$$

$$H_L = a_L \left(\frac{W_L}{\mu} \right)^d Sc_L^{0.5}, \tag{6.33}$$

where W_L and W_G are the fluxes in $lb_m/ft^2 \cdot hr$ ($kg/m^2 \cdot hr$), Sc_G and Sc_L are the Schmidt numbers for the gas and liquid phases, respectively, and μ is the liquid-phase viscosity in units of $lb_m/ft \cdot hr$ ($kg/m \cdot hr$). The values of H_L and H_G are computed in units of ft (m). The constants are given in Table 6.4. The Schmidt number is calculated from $Sc = \mu/\rho D_{AB}$ where ρ is density and D_{AB} is the diffusion coefficient of A diffusing through B.

Table 6.3 *Definition for the number of transfer units and the height of a transfer unit [5]. Reprinted by permission of John Wiley and Sons, Inc. Copyright © 1980.*

Mechanism	Driving force		Number of transfer units (NTU)		Height of a transfer unit (HTU)
Equimolar counter diffusion	$y_{A_I} - y_A$	N_G	$\displaystyle\int_{y_{A_{out}}}^{y_{A_{in}}} \frac{dy_A}{y_{A_I} - y_A}$	H_G	$\displaystyle\frac{V}{k_y a A_c}$
	$y_A^* - y_A$	N_{OG}	$\displaystyle\int_{y_{A_{out}}}^{y_{A_{in}}} \frac{dy_A}{y_A^* - y_A}$	H_{OG}	$\displaystyle\frac{V}{K_y a A_c}$
	$x_A - x_{A_I}$	N_L	$\displaystyle\int_{x_{A_{in}}}^{x_{A_{out}}} \frac{dx_A}{x_A - x_{A_I}}$	H_L	$\displaystyle\frac{L}{k_x a A_c}$
	$x_A - x_A^*$	N_{OL}	$\displaystyle\int_{x_{A_{in}}}^{x_{A_{out}}} \frac{dx_A}{x_A - x_A^*}$	H_{OL}	$\displaystyle\frac{L}{K_x a A_c}$
Diffusion through a stationary component	$y_{A_I} - y_A$	N_G	$\displaystyle\int_{y_{A_{in}}}^{y_{A_{out}}} \frac{(1 - y_A)_{\ell m} dy_A}{(1 - y_A)(y_{A_I} - y_A)}$	H_G	$\displaystyle\frac{V}{k_y' a A_c(1 - y)_{\ell m}}$
	$y_A^* - y_A$	N_{OG}	$\displaystyle\int_{y_{A_{in}}}^{y_{A_{out}}} \frac{(1 - y_A)_{\ell m} dy_A}{(1 - y_A)(y_A^* - y_A)}$	H_{OG}	$\displaystyle\frac{V}{K_y' a A_c(1 - y)_{\ell m}}$
	$x_A - x_{A_I}$	N_L	$\displaystyle\int_{x_{A_{in}}}^{x_{A_{out}}} \frac{(1 - x_A)_{\ell m} dx_A}{(1 - x_A)(x_A - x_{A_I})}$	H_L	$\displaystyle\frac{L}{k_x' a A_c(1 - x)_{\ell m}}$
	$x_A - x_A^*$	N_{OL}	$\displaystyle\int_{x_1}^{x_2} \frac{(1 - x_A)_{\ell m} dx}{(1 - x_A)(x - x_A^*)}$	H_{OL}	$\displaystyle\frac{L}{K_x' a A_c(1 - x)_{\ell m}}$

a = specific surface of column packing
A_c = empty tower cross-sectional area

Table 6.4 *Constants for determining* H_G *and* H_L. *Range of* WL_L *for* H_L *is 400 to 15,000 [6].*

Packing (in)	a_G	b	c	a_L	d	Range for H_G W_G (lb$_m$/ft$^2 \cdot$ hr)	W_L (lb$_m$/ft$^2 \cdot$ hr)
Raschig rings							
$\frac{3}{8}$	2.32	0.45	0.47	0.0018	0.46	200–500	200–500
1	7.00	0.39	0.58	0.010	0.22	200–800	400–500
1	6.41	0.32	0.51	–	–	200–600	500–4500
Berl saddles							
$\frac{1}{2}$	32.4	0.30	0.74	0.0067	0.28	200–700	500–1500
$\frac{1}{2}$	0.811	0.30	0.24	–	–	200–700	1500–4500
1	1.97	0.36	0.40	0.0059	0.28	200–800	400–4500

Example 6.5: packed column

Problem:

A gas mixture of 2 mol% SO_2 (*A*) and 98 mol% dry air (*B*) is to be treated with pure water in a packed absorption column using 1-inch Berl saddles as packing. The exit gas should contain 0.1 mol% or less SO_2. The column operates at 30 °C and 1 atm pressure. The gas and liquid fluxes in the column are $W_G = 250$ lb$_m$/ft$^2 \cdot$ hr and $W_L = 4000$ lb$_m$/ft$^2 \cdot$ hr. Determine the required column height, given:

Air	**Water**
$\mu = 0.07$ kg/m $\cdot$ hr	$\mu = 3.6$ kg/m $\cdot$ hr $= 2.5$ lb$_m$/ft $\cdot$ hr
$D_{AB} = 0.046$ m^2/hr	$D_{AB} = 6.1 \times 10^{-6}$ m^2/hr
$\rho = 1.16$ kg/m^3	$\rho = 997$ kg/m^3.

VLE equation: $y = 26x$.

Solution:

$$Sc_G = \left(\frac{\mu}{\rho D}\right)_G$$

$$= \frac{0.07 \text{ kg/m} \cdot \text{hr}}{\left(1.16 \text{ kg/m}^3\right)\left(0.046 \text{ m}^2/\text{hr}\right)} = 1.3.$$

$$Sc_L = \left(\frac{\mu}{\rho D}\right)_L$$

$$= \frac{3.60 \text{ kg/m} \cdot \text{hr}}{\left(997 \text{ kg/m}^3\right)\left(6.1 \times 10^{-6} \text{ m}^2/\text{hr}\right)} = 592.$$

Using constants for 1-inch Berl saddles:

$$H_G = 1.97(250)^{0.36}(1.3)^{0.5}/(4000)^{0.4}$$

$$= 0.6\,\text{ft}.$$

$$H_L = 5.9 \times 10^{-3}(592)^{0.5}\left(\frac{4000}{2.5}\right)^{0.28}$$

$$= 1.13\,\text{ft}.$$

$$H_{OG} = H_G + \frac{mV}{L}\,H_L$$

$$= 0.6\,\text{ft} + \frac{26(250)}{4000}(1.13\,\text{ft})$$

$$= 2.44\,\text{ft}.$$

$$A = \frac{mV}{L}$$

$$= \frac{26(250)}{4000} = 1.63.$$

$$x_{A_{in}} = 0 \qquad y^*_{A_{out}} = mx_{A_{in}} = 0$$

$$N_{OG} = \frac{\ln\left\{\left[\dfrac{A-1}{A}\right]\left[\dfrac{y_{A_{in}} - mx_{A_{in}}}{y_{A_{out}} - mx_{A_{in}}}\right] + \dfrac{1}{A}\right\}}{\left(\dfrac{A-1}{A}\right)}$$

$$= \ln\frac{\left\{\left[\dfrac{0.63}{1.63}\right]\left[\dfrac{0.02}{0.001}\right] + \dfrac{1}{1.63}\right\}}{\dfrac{0.63}{1.63}}$$

$$= 5.45.$$

Column height, $\quad z = H_{OG}N_{OG}$

$$= (2.44\,\text{ft})(5.45) = 13.3\,\text{ft}.$$

Example 6.6: packed column absorption

Problem:

Air at 25 °C is used to dry a plastic sheet containing acetone. At the drier exit, the air leaves containing 0.02 mole fraction acetone. The acetone is to be recovered by absorption with water in a packed tower. The gas composition is to be reduced to 5×10^{-3} mole fraction at the column exit. The equilibrium relationship is $y = 1.8x$. The gas enters the bottom of the column (Figure 6.12) at a flux of 1000 $\text{lb}_m/\text{ft}^2 \cdot \text{hr}$, and the water enters the top at a flux of 1400 $\text{lb}_m/\text{ft}^2 \cdot \text{hr}$. The tower is packed with 1-inch Berl saddles. Calculate the column height, given:

Water

$Sc_L = 915$

$\mu = 12\,\text{lb}_m/\text{ft} \cdot \text{hr}$

$$\text{Molar flux of gas} = \frac{1000 \text{ lb}_m/\text{ft}^2 \cdot \text{hr}}{0.02 \times 58 + 0.98 \times 29} = 34 \frac{\text{lb}_m}{\text{ft}^2 \cdot \text{hr}} = W'_G.$$

Acetone in air

$Sc_G = 1.7.$

Solution:

Molecular weight of acetone $= 58$

Molecular weight of air $= 29$

Molecular weight of water $= 18$

$$\text{Molar liquid flux} = \frac{1400 \text{ lb}_m/\text{ft}^2 \cdot \text{hr}}{18} = 78 \frac{\text{lb}_m}{\text{ft}^2 \cdot \text{hr}} = W'_L.$$

Steady-state mass balance

$$y_{\text{in}} W'_G = y_{\text{out}} W'_G + x_{\text{out}} W'_L$$

$$(0.02)(34) = (0.005)(34) + x_{\text{out}}(78)$$

$$x_{\text{out}} = 6.5 \times 10^{-3}$$

$$A = \frac{mV}{L} = \frac{mW'_G}{W'_L} = \frac{1.8(34)}{(78)} = 0.79.$$

Both operating and equilibrium lines are linear:

$$N_{OG} = \frac{\ln\left\{\dfrac{A-1}{A}\left[\dfrac{y_{A_{\text{in}}} - mx_{A_{\text{in}}}}{y_{A_{\text{out}}} - mx_{A_{\text{in}}}}\right] + \dfrac{1}{A}\right\}}{\left(\dfrac{A-1}{A}\right)}$$

$$= 1.5$$

$$H_{OG} = H_G + \frac{mV}{L} H_L$$

$$H_L = a_L \left(\frac{W_L}{\mu}\right)^d (Sc_L)^{0.5} = 0.0059 \left(\frac{1400}{12}\right)^{0.28} (915)^{0.5} = 0.7 \text{ ft}.$$

$$H_G = a_G W_G^b Sc_v^{0.5}/W_L^c$$

$$= 1.97(1000)^{0.36}(1.7)^{0.5}/(1400)^{0.24}$$

$$= 0.5 \text{ ft}.$$

$$H_{OG} = H_G + \frac{mV}{L} H_L = 0.5 + 0.79(0.7) = 1.05 \text{ ft}.$$

Column height, $z = H_{OG} N_{OG} = (1.05 \text{ ft})(1.5) = 1.6 \text{ ft}.$

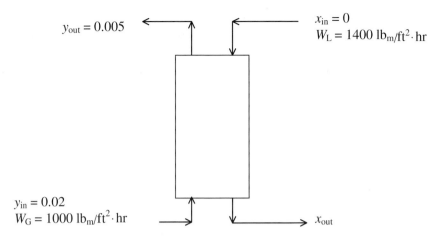

$y_{out} = 0.005$

$x_{in} = 0$
$W_L = 1400 \text{ lb}_m/\text{ft}^2 \cdot \text{hr}$

$y_{in} = 0.02$
$W_G = 1000 \text{ lb}_m/\text{ft}^2 \cdot \text{hr}$

x_{out}

Figure 6.12 Schematic for air purification, Example 6.6.

6.7   Remember

- Absorption and stripping are UNIT OPERATIONS. Regardless of what chemicals are being separated, the basic design principles for absorption and stripping are always similar.
- The assumption that stages in an absorption or stripping column are in equilibrium allows calculations of concentrations without detailed knowledge of flow patterns and mass transfer rates. This assumption is a major simplification.
- Because one feed stream is already a vapor phase, absorption or stripping columns usually do not require condensers and reboilers.
- The McCabe–Thiele analysis can be used to model physical absorption and stripping processes that use equilibrium stages.
- The HTU–NTU method can be used to model absorption or stripping columns which contain continuous packing instead of equilibrium stages.

6.8   Questions

6.1 Explain, in terms of Henry's Law coefficients, whether or not increasing the temperature in a stripping column improves the separation at each stage.

6.2 Explain the tradeoff between high and low liquid velocities in absorption in terms of column operation and effective mass transfer between the phases.

6.3 It is typical to assume that both absorption and stripping columns operate at constant pressure. How would non-constant pressure affect the design? Consider pressure either increasing or decreasing down the length of the column.

6.4 Show graphically how the slope of the operating line affects the degree of separation that can occur in an absorption column.

6.5 Under which circumstances does the gas phase individual mass transfer coefficient dominate a separation? When is the liquid phase individual mass transfer coefficient dominant?

6.6 As the contaminant concentration in a system increases, what happens to the assumption of equimolar counter transfer?

6.7 The statement has been made that "one can show that each ten-fold reduction in concentration is equally difficult" for gas absorption with dilute solutions. Is this reasonable?

6.9 Problems

6.1 Prove that for *dilute* systems: $(1 - y_A)_{\ell m} = 1$, and $k'_y a = k_y a$. What does this mean?

6.2 Carbon disulfide, CS_2, used as a solvent in a chemical plant, is evaporated from the product in a drier to an inert gas (essentially N_2) in order to avoid an explosion hazard. The vapor–N_2 mixture is to be scrubbed with an absorbent oil, which will be subsequently steam stripped to recover the CS_2. The CS_2–N_2 mixture has a partial pressure of CS_2 equal to 50 mmHg at 24 °C (75 °F) and is to be blown into the absorber at essentially standard atmospheric pressure at the expected flowrate of 0.40 m^3/s (50,000 ft^3/hr). The vapor content of the gas is to be reduced to 0.5%. The absorption oil has an average molecular weight of 180, a viscosity of 2 cP, and a specific gravity of 0.81, at 24 °C. The oil enters the absorber essentially stripped of all CS_2, and solutions of CS_2 and oil, while not actually ideal, follow Raoult's Law. The vapor pressure of CS_2 at 24 °C is 346 mmHg. Assuming isothermal operation [8]:

(a) Determine the minimum liquid/gas ratio

(b) A tray tower containing six equilibrium stages is available for this separation. What oil flowrate is necessary to accomplish this? Solve this part graphically.

6.3 A gas stream is 90 mol% N_2 and 10 mol% CO_2. The plan is to absorb the CO_2 into pure water at 5 °C. Assume isothermal operation at 10 atm and a liquid flowrate of 1.5 times the minimum. How many equilibrium stages are required to absorb 90% of the CO_2?

7

Adsorption

There is a continual exchange of ideas between all minds of a generation.

– AUGUSTE RODIN (1911)

7.1 Objectives

1 Define the concepts of mass transfer zone, breakthrough, and exhaustion.
2 Use the scale-up approach and the kinetic approach to design fixed-bed adsorption columns based on laboratory or pilot column data.

7.2 Background

Adsorption is a process whereby a substance (adsorbate, or sorbate) is accumulated on the surface of a solid (adsorbent, or sorbent). The adsorbate can be in a gas or liquid phase. The driving force for adsorption is unsaturated forces at the solid surface which can form bonds with the adsorbate. These forces are typically electrostatic or van der Waals interactions (reversible). Stronger interactions involve direct electron transfer between the sorbate and the sorbent (irreversible). The strength of this interaction dictates the relative ease or difficulty in removing (desorbing) the adsorbate for adsorbent regeneration and adsorbate recovery. The selective nature of the adsorbent is primarily due to the relative access and strength of the surface interaction for one component in a feed mixture. The solid is the mass-separating agent and the separating mechanism is the partitioning between the fluid and solid phases. An energy-separating agent, typically a pressure or temperature change, is used to reverse the process and regenerate the sorbent.

Adsorption processes are used economically in a wide variety of separations in the chemical process industries. Activated carbon is the most common adsorbent, with annual worldwide sales estimated at $380 million [1]. One common adsorption process is dehydration for the drying of gas streams.

Adsorption offers several advantages as a separation process. It can be used in situations where distillation is difficult or impossible due to components with similar boiling points, vapor–liquid azeotropes, or species with low relative volatilities. A high loading of solute is possible in adsorption processes and it works well for dilute systems. In addition the energy-separating agent needs are usually low. The disadvantages of adsorption processes are due to the use of an adsorbent mass-separating agent. Each adsorbent bed must be removed from the process before regeneration, so that more than one column in series or parallel is usually required. The regeneration process can involve losses in the sorbent amount and loading over time. Column operation affects solute breakthrough that limits actual bed loading.

7.2.1 Definition of adsorption terms

- Adsorption: the process where a substance is accumulated on an interface between phases.
- Adsorbent: the phase that collects the substance to be removed at its surface.
- Adsorbate: the solute that is to be adsorbed (removed from the gas or liquid stream).
- Isotherm: A relation between the equilibrium amount of a substance adsorbed per weight of sorbent and its concentration in the liquid or gas stream at constant temperature.

7.2.2 Environmental applications

There are many environmental applications of adsorption in practice and many others are being developed. Both zeolite and activated carbon adsorbents are used in VOC removal from gas streams. Molecular sieves are used to remove water from organic solvents, while other adsorbent materials remove organics from water. Taste and odor and other contaminant removal in water treatment is performed with activated carbon and other adsorbents. Silica gel, activated carbon, zeolites, activated alumina, and synthetic resins have all been applied to removal of H_2S from gas streams. Phillips Petroleum has patented a z-sorb technology that removes H_2S and other sulfur compounds from gas streams above 600 °F (316 °C) [2]. Adsorption is used to eliminate purge streams to remove contaminants. Mercury can be removed from chlor–alkali-cell gas effluent via adsorption. Other adsorbents, such as bentonite [3], are studied as heavy-metal adsorbents in clay barriers. Water removal from gas streams containing acid gases, odor or contaminant removal from air, and radon removal from gas streams are all achieved through adsorption. In addition, adsorption can be used to eliminate solvent use as an alternative to extraction or azeotropic distillation.

7.3 Adsorption principles

7.3.1 Physical vs chemical adsorption

Physical adsorption occurs due to van der Waals (dispersion) or electrostatic forces. The attraction depends on the polar nature of the fluid component being adsorbed as well as that of the adsorbent. Van der Waals forces are directly related to the polarizability. An estimate of the relative strength of interaction is based on the sorbate size and polarizability. Electrostatic forces include polarization forces, field–dipole interactions and field gradient–quadrupole interactions. These forces arise when the surface is polar. In the case of a polar solvent like water with non-polar organic impurities, the organic molecules will prefer to stick to a non-polar adsorbent such as activated carbon rather than remain in the polar solvent. Physical adsorption is reversible. Physical sorption is sensitive to temperature, relatively non-specific regarding sorbates, relatively fast kinetically, and has a low heat of adsorption ($<2\Delta H_{vap}$). Multiple sorbate layers can form on the sorbent surface.

Chemical adsorption (chemisorption) occurs when the attraction between the adsorbent and the adsorbate can form a covalent bond, or when a chemical reaction occurs between the adsorbent and adsorbate. Usually chemical adsorption will only allow a single layer of molecules (monolayer) to accumulate on the surface of the adsorbent. Chemical adsorption is usually irreversible. Chemisorption is typically more specific, kinetically slower and has a larger heat of adsorption ($>3\Delta H_{vap}$).

7.3.2 Separating mechanism

The interaction of the adsorbate with the solid surface can be due to three mechanisms: steric, kinetic, or equilibrium. Steric interactions are due to the shape of the molecule. One example would be the difference in adsorption strength (heat of sorption) for a linear vs a branched hydrocarbon. Another example would be the separation of a large and small molecule using zeolites whereby the small molecule could enter the zeolite pores and the large molecule would be excluded (molecular sieving). Kinetic interactions are due to the relative ease of accessibility of the adsorbate to the solid surface. Diffusion through the fluid boundary layer to the solid surface and diffusion in the pores of the sorbent contribute to this effect (i.e., whichever component gets to the sorption site first wins). Equilibrium interactions relate to the thermodynamic equilibrium state of the fluid and solid phases. Equilibrium interactions differ from kinetic effects in that a molecule may get to the sorption site first (kinetic) but will later be displaced by the more strongly adsorbed species due to the reversibility of the process. For physical sorption, there is a finite rate of adsorption as well as desorption to the solid surface. When these rates are equal, the process is at equilibrium. This is analogous to a reversible chemical reaction where the equilibrium constant represents the balance of the forward and reverse reaction rates.

7.3.3 Criteria for use

Distillation is usually the first choice for bulk separation of components in a liquid phase. Another option often used is air stripping. In both cases, the ease of separation is based on the relative volatility of the various components in the feed mixture. In comparison, adsorption is a good choice when [4]:

1 The relative volatility between the key components to be separated is in the range of 1.2 to 1.5 or less. In this case, distillation is not an obvious choice and the separation factor for adsorption could be very large based on criteria other than relative volatility.

2 The bulk of the feed is a relatively low value, more volatile component, and the product of interest is in relatively low concentration. For this case, large reflux ratios (and large energy consumption) would be required if distillation is considered. This situation is often found in environmental applications where one has dilute concentrations of a pollutant in an air or water stream.

3 The two groups of components to be separated have overlapping boiling ranges. Again, compared to distillation, several distillation columns are required even if the relative volatilities are large. For adsorption, this separation can be done effectively if the two groups contain chemically or geometrically dissimilar molecules so as to obtain a high separation factor.

4 A low temperature and a high pressure are required for liquefaction (for gases).

5 Components in the feed stream can be damaged or altered by heating. These alterations can include chemical degradation, increases in chemical reactivity and polymerization reactions. This chemical change can cause fouling in the equipment which reduces the effective lifetime as well as the production of undesirable components.

6 Chemical corrosion, precipitation, and/or undesirable chemical reactions (explosive conditions are an example) are problems using operating conditions for distillation.

7 Factors favorable for separation by adsorption exist. Adsorption selectivity between the key components should be greater than 2, the sorbent can be easily regenerated, does not easily foul for the given feed mixtures, and does not act as a catalyst or other reaction medium to promote the production of undesirable side products. The feed throughput and product purity are also important considerations. The costs for adsorption separation are generally lower than distillation for small to medium throughputs (less than a few tons per day), and when high-purity products are not required (function of selectivity of the sorbent with respect to the feed components). The major cost for pressure-swing adsorption (discussed in the next section) is the compressor costs. If a feed gas mixture is available at an elevated pressure, the separation costs are substantially reduced.

7.4   Sorbent selection

Since almost all adsorptive separation processes are based on equilibrium partitioning, the most important factor to consider initially is the adsorption isotherm (equilibrium

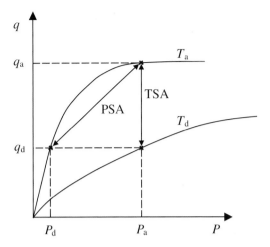

Figure 7.1 Isotherm diagram illustrating pressure-swing (PSA) and temperature-swing (TSA) processes [5]. Reproduced with permission of the American Institute of Chemical Engineers. Copyright © 1988, AIChE. All rights reserved.

relationship). The isotherm should span the concentration range required. If there is a temperature change, then this information also needs to be available over the temperature range of interest. These data are needed at both the adsorption and desorption (sorbent regeneration) conditions. It is usually necessary to change the operating conditions of the adsorption process during regeneration to make desorption more thermodynamically favorable. This can involve an increase in temperature and/or a reduction of partial pressure, and data at these new conditions can be helpful in the analysis.

Figure 7.1 illustrates the data required for the various sorption options. The sorption isotherm is plotted for two temperatures. For a pressure-swing process (PSA), the adsorption step is performed at P_a and the desorption at P_d. The maximum amount of solute removed and recovered per mass of sorbent is $q_a - q_d$. The temperature-swing process (TSA) is also illustrated. A combination of temperature and pressure swings can be used, although this is rarely done in practice. Note that the absolute values of the pressure and temperature as well as the change (swing) affect the productivity of the process. In addition, pressure changes can be accomplished very rapidly while temperature changes occur much more slowly since the entire bed must be heated or cooled. Typical cycle times for PSA are 1–5 minutes while it is 2 hours or longer for TSA. Once the sorption isotherm information is obtained, then other factors need to be considered in adsorption design [5]:

1 Sorbent capacity (i.e., how much material is adsorbed per unit quantity of sorbent). The surface area per unit volume of the sorbent has an important influence on this value. For this reason, sorbents are usually highly porous materials. Obviously, one would want the capacity to be as high as possible.
2 The purity requirement of the fluid phase (i.e., how much material needs to be removed).
3 Sorbent selectivity. This can be accomplished by three mechanisms: (a) selective binding to the sorbent surface (equilibrium); (b) excluding certain components based

on geometric factors (sieving); and, (c) differences in diffusion rates from the fluid to solid surface, usually the intra-particle diffusion rate (kinetic). Most sorbents are based on (a).

4 The method of sorbent regeneration (see below).

5 The sorbent bed volume. This value includes the portion of the bed which is not completely saturated when the sorption step is stopped as well as the void space in the bed. The mass transfer rates (transport processes) for sorption have a large influence on the bed size and dimensions. Some design approaches are based on sizing the unused portion of the sorbent bed, as discussed in Section 7.8.

6 Sorbent deactivation. Components in the feed stream can strongly adsorb and/or react with the sorbent and substantially reduce its effective lifetime.

7 Cost. Less expensive sorbents that can be easily regenerated are preferred.

The type of sorbent that will work for a given application will be chosen by consideration of these factors. Once the sorbent type is chosen, the particle size and shape need to be considered. These factors affect both the pressure drop in the sorbent bed and the mass transfer rate from the fluid to the sorbent surface. As general rules of thumb, the pressure drop through the bed increases as the particle size decreases. In contrast, the mass transfer rate usually increases as the particle size decreases. The factors do not vary to the same degree with particle size so some balance needs to be made in the chosen design [6].

7.5 Various sorbents

Table 7.1 lists the typical sorbents used; their uses as well as strengths and weaknesses. The four major commercial adsorbents are the following: zeolite molecular sieves (zms), activated alumina, silica gel, and activated carbon. The surfaces of activated alumina and most molecular-sieve zeolites are hydrophilic, and will preferentially adsorb water over organic molecules. Silicalite, which is a hydrophobic zeolite, is the main exception. Activated carbon, on the other hand, preferentially adsorbs organic and non-polar or weakly polar compounds over water. The surface of silica gel is somewhere in between these limits and has affinity for both water and organics. Detailed information about each of these classes of adsorbents can be found in Refs. [1, 4, 6, 7].

7.5.1 Activated carbon

Activated carbon is prepared by (a) heating organic materials such as wood, coal char, almond, coconut, or walnut shells, as well as vinyl copolymers or recycled tires in a stoichiometric O_2-deficient atmosphere, and (b) activating the product by exposure to a mild oxidizing gas (CO_2 or steam, for example) at a high temperature. The activation

Table 7.1 *Adsorbent types [1]. Reproduced with permission of the American Institute of Chemical Engineers. Copyright © 1988, AIChE. All rights reserved.*

Adsorbent	Characteristics	Commercial uses	Strengths	Weaknesses
Activated carbon	Hydrophobic surface, favors organics over air or water	Removal of organic pollutants from aqueous and gaseous effluents	Cheapest hydrophobic adsorbent, workhorse of the pollution-control business	Difficult to regenerate if fouling occurs, sometimes can catch fire during regeneration
Carbon molecular sieves	Separates on the basis of different intra-particle diffusivities	Production of N_2 from air	The only practical adsorbent favoring O_2 adsorption over N_2	Has not found any uses except for air separation
Silica gel	High-capacity, hydrophilic adsorbent	Primarily drying of gas streams, sometimes used for hydrocarbon removal from gases	Higher capacity than zeolite molecular sieves (zms)	Not as effective as zms in removing traces of H_2O from gases
Activated alumina	High-capacity hydrophilic adsorbent	Primarily drying of gas streams	Higher capacity than zms	Not as effective as zms in removing traces of H_2O from gases
Zeolite molecular sieves (zms)	Hydrophilic surface; polar, regular chemicals	Dehydration, air separation, geometry-based separations, many others	Can make separations based both on polarity and geometry	Lower total capacity than many other adsorbents
Silicalite	Hydrophobic surface; adsorption characteristics similar to those of activated carbon	Removal of organics from gas streams	Can be burned off more easily than activated carbon	More costly than activated carbon
Polymer adsorbents	Usually styrene	Removal of organics from gas streams	Not as subject to fouling as activated carbon	Much more costly than activated carbon
Irreversible adsorbents	Surfaces that react selectively with components of gas streams	Removal of low levels of H_2S, SO_2, etc., from gases	Excellent in removing trace contaminates	Only economical for removal of <100 kg/day of adsorbents
Biosorbents	Activated sludge on a porous support	Removal of organics from gas streams	No regeneration needed	Percent removals often lower than for other adsorbents

step creates the pore structure, and the surface area of the particle is greatly increased. The activation conditions are varied to obtain the desired pore structure and mechanical strength.

Activated carbon is the most common adsorbent due to its large surface area per unit mass (300 to 1,500 m^2/g). The surface area per unit volume and the pore-size distribution vary depending on whether the application is for liquid- or gas-phase feed streams. Larger pore sizes are used for liquid-phase streams (30-Å-diameter, as opposed to the 10–25-Å-diameter carbons used for gas-phase feeds) due to the larger size of the sorbates and the slower diffusion rates for liquids.

The surface of activated carbon is non-polar or only slightly polar as a result of the surface oxide groups and inorganic impurities. Most other commercially available sorbents are polar in nature. This difference has some very useful advantages. Activated carbon does not adsorb water very well. So, it does not require any pretreatment to remove water prior to use and is very useful as a selective sorbent for aqueous systems (aquarium filters!). This property also makes it a useful sorbent for non-polar or weakly polar sorbates: one environmental application of activated carbon is its use in canisters on automobile fuel tanks to reduce hydrocarbon emissions (from volatilization). VOC removal from water or gas streams is another large application. The strength of the sorbate–sorbent interaction (heat of adsorption) is generally lower for activated carbon. Desorption and sorbent regeneration are thus easier and require less energy than other sorbents. Granular activated carbon can be regenerated by heating to oxidize the collected organic matter. Usually about 5–10% of the original amount is lost, and the regenerated carbon loses some of its adsorptive capacity. The same process is not economically feasible for PAC.

Activated carbon is available as 1–3-mm-diameter beads and 2–4-mm-diameter pellets. It is also available in granular (GAC) and powdered (PAC) forms. The granular form is often used in beds and columns. The powdered form is more likely to be added to a stirred-tank reactor, and settled or filtered out after the adsorption process is considered to be complete.

7.5.2   Activated alumina

High surface area per unit mass alumina, either amorphous or crystalline, which has been partially or completely dehydrated is termed activated alumina. This material is very hydrophilic and is often used for the drying or dehydration of gases and liquids. Environmental applications would include water removal from acid gas or organic solvent streams. Activated alumina is produced by thermal dehydration or activation of $Al_2O_3 \cdot nH_2O$ ($n = 1, 3$) to get n close to 0.5. The surface area per unit mass of this material is usually in the range of 200–400 m^2/g. The predominant pore diameters are in the 2–5-nm range. It can be obtained as spheres (1–8-mm diameter), pellets (2–4-mm diameter), granules, and powder.

7.5.3 Silica gel

Silica gel is one of the synthetic amorphous silicas. It is a rigid network of spherical colloidal silica particles. It is often sold in two forms: regular density, which has a surface area per unit mass range of 750–850 m^2/g (average pore diameter = 22–26 Å); and low density, which has a surface area per unit mass of 300–350 m^2/g (average pore diameter = 100–150 Å). Silica gel is prepared by mixing a sodium silicate solution with a mineral acid such as sulfuric or hydrochloric acid. The reaction produces a concentrated dispersion of finely divided particles of hydrated SiO_2, known as silica hydrosol or silicic acid. The hydrosol, on standing, polymerizes into a white jelly-like precipitate, which is silica gel. This gel is washed, dried and activated. Properties such as surface area per unit mass, pore volume, and strength are varied by adjusting reaction conditions.

Silica gel is used for water removal applications. Regeneration is achieved by heating to approximately 150 °C, as compared to 350 °C for zeolites, where the heats of adsorption for water are considerably higher. Zeolites, however, have the advantage of higher water capacities at low relative pressures; hence they are used at high temperatures.

7.5.4 Zeolites

Zeolites are nanoporous oxide crystalline structures, typically aluminosilicates. The aluminum in the structure has a negative charge that must be balanced by a cation, M. This ionic structure leads to the hydrophilicity of the zeolite. Silicalite, a pure silica version, is charge neutral and hydrophobic. Zeolites have uniform pore sizes that typically range from 0.3 to 0.8 nm. The pore size and/or adsorption strength can be altered by the type and number of cations present in the structure. The void fraction can be as high as 0.5. Zeolites can selectively adsorb or reject molecules based on their size, shape, or sorption strength. The molecular sieving effect is a common term associated with zeolites, and refers to selectivity based on size or shape exclusion. Zeolites can also provide separations based on competitive sorption. This situation can lead to reverse selectivity where a larger molecule can be selectively sorbed and separated from a smaller molecule. For example, most zeolites are polar adsorbents and will preferentially adsorb polar species (i.e., water) over non-polar species (organics) of comparable size.

Separation can be based on the molecular-sieve effect and/or selective adsorption. These separations are governed by several factors [7]:

1 The basic framework structure of the zeolite determines the pore size and the void volume.

2 The exchange cations, in terms of their specific location in the structure, number density, charge, and size, affect the molecular-sieve behavior and adsorption selectivity of the

Table 7.2 *Zeolite molecular-sieve adsorbents [8]. Reproduced with permission of the American Institute of Chemical Engineers. Copyright © 1987, AIChE. All rights reserved.*

Designation	Alternative name	Channel diameter, nm	Application
KA	3A	0.29	Drying of various gases
NaA	4A	0.38	CO_2 removal from natural gas
CaX	–	0.80	Removal of mercaptans from natural gas
Mordenite	–	0.70	I and Kr removal from nuclear off-gases
Silicalite	–	0.55	Removal of organics from water

zeolite. By changing the cation type and number, one can modify the selectivity of the zeolite for a given separation.

3 The effect of the temperature can be substantial in situations involving activated diffusion.

Many zeolites occur naturally, but the majority of those used commercially have been synthesized and are designated by a letter or group of letters (Type A, Type Y, Type ZSM, etc.). The lettering system has evolved empirically and has no relation to the structure. For example, ZSM stands for Zeolite Sacony Mobil.

Tables 7.2, 7.3, and 7.4 list some zeolites and their environmental applications.

7.6   Sorbent regeneration

In most instances, it is necessary to regenerate the sorbent after each cycle of use. This step is important to reduce the replacement costs of the sorbent, extend the useful life of the sorbent and minimize the quantity of solid material that must be discarded. For gases, there are two methods generally used. In pressure-swing adsorption (PSA), the feed is introduced at a high pressure to obtain a high degree of solute adsorption. When the sorbent bed is saturated, the pressure is reduced to remove the sorbed species for recovery or disposal and regenerate the sorbent. The pressure range will be dictated by cost and the sorption isotherm. The pressure swing can be accomplished very rapidly. Temperature-swing adsorption (TSA) uses heat to regenerate the sorbent. The feed is introduced at a low temperature for sorption. Heat is applied to the saturated bed to provide sufficient thermal energy to desorb the sorbates and regenerate the sorbent. This temperature swing can take several hours since it is necessary to heat the entire bed. This procedure is only used when the bed volumes are small and the components are not damaged by the temperature change. A combination of increased temperature and reduced pressure may also be used during sorbent regeneration.

Table 7.3 *Environmental applications of zeolites [9]. Reproduced with permission of the American Institute of Chemical Engineers. Copyright © 1999, AIChE. All rights reserved.*

Application	Zeolites used	Advantages
Selective catalytic reduction of NO_x	Copper ZSM-5; mordenite	Good for high dust applications Extended temperature range Cheaper Higher selectivity Mordenite is particularly stable in acid streams
Lean NO_x	Copper, cobalt ZSM-5; beta	Uses fuel hydrocarbons as reductants No ammonia No special handling Cheaper
Lean-burn (oxygen-rich) diesel-engine NO_x removal	Copper, cobalt ZSM-5; beta	More effective than three-way catalytic converter for NO_x
Removal of N_2O	Cobalt, copper ZSM-5; mordenite; ferrierite; beta; ZSM-11	N_2O decomposes over zeolites at higher temperatures (400 °C)
VOC removal in dilute, high-volume, humid streams	High-silica, hydrophobic zeolites	Effective where carbon is not Systems available from several vendors
VOC removal during automotive cold starts	High-silica, hydrophobic medium- and large-pore zeolites	Achieved 35–70% reduction

Two additional methods which can shift equilibrium conditions to favor desorption are purge stripping and displacement, as described by Ruthven [5].

In a thermal swing cycle, the bed may be purged either with feed gas or with an inert gas. The inert purge has the advantage in that the theoretical purge volume required to clean the bed is reduced. Since the bed is generally also heated by the purge, however, the purge requirement may in practice be determined by the heat balance, not by equilibrium considerations. The theoretical advantage may not, therefore, be realizable and purging with hot feed may prove more economical. Isothermal purge gas stripping is seldom economical since an inordinately large purge volume would be required. However, displacement desorption, in which a competitively adsorbed species is used to displace the strongly adsorbed feed component, is commonly employed for systems in which thermal swing operating is precluded, for example by the reactivity of the sorbate. Since the displacement cycle requires the displacing agent to be recovered and recycled, it is used generally only when simpler cycles are impractical. Steam stripping, which is commonly used for regeneration of activated carbon beds, may be regarded as a combined displacement/thermal swing operation.

Table 7.4 *Waste reduction through process improvements using zeolites [9].*
Reproduced with permission of the American Institute of Chemical Engineers.
Copyright © 1999, AIChE. All rights reserved.

Application	Zeolites used	Advantages
Production of alpha-terpinyl alkyl ethers	Beta	Excellent yields in continuous reactor Eliminates use of HCl, H_2SO_4, $AlCl_3$, toluene, sulfonic acid, boron trifluoride etherate, and acidic cation resins as catalysts
Cumene synthesis	Dealuminated mordenite; MCM-22; beta; Y; omega	Lower impurities Transalkylation function Lower benzene-to-propylene ratio allows higher capacity, great unit efficiency High selectivity Regenerable, non-hazardous, non-corrosive
Direct oxidation of benzene to phenol	ZSM-5	Eliminates cumene as an intermediate Enables possible use of N_2O as oxidant
Caprolactam (via oxidation)	Titanium-framework-substituted ZSM-5 (TS-1)	Dramatic reduction in number of processing steps and waste streams Possible further reduction by using another zeolite in the last step of the process
Gasoline from methanol	ZSM-5	Produces methanol from coal, natural gas, or biomass and then converts it into liquid fuel Conservation of crude oil, elimination of many waste streams

The regenerability of an adsorbent determines the fraction of capacity, or working capacity, that is recovered for future use. In most cases, a constant decrease in working capacity occurs after the first cycle and is maintained for up to approximately 50–100 cycles. Eventually, however, slow aging or gradual poisoning causes the working capacity to be reduced to the point that the adsorbent needs to be replaced.

For liquid feeds, the situation is changing. Historically, sorbents would be used for only one cycle and then discarded. The costs of this disposal are rising very rapidly and can usually only be justified for the processing of small fluid volumes. For regeneration, as much liquid as possible is drained from the sorbent initially. If the sorbate is volatile, the sorbent bed can be heated for the desorption step and the sorbate recovered by condensation. A sweep gas can be used to assist in the removal. Another possibility is to pass a second fluid through the bed that has a high partition coefficient for the sorbate. This approach

Table 7.5 *Factors governing choice of regeneration method [5]. Reproduced with permission of the American Institute of Chemical Engineers. Copyright © 1988, AIChE. All rights reserved.*

Method	Advantages	Disadvantages
Thermal swing	Good for strongly adsorbed species: small change in T gives large change in q (heat of sorption) Desorbate may be recovered at high concentration Gases and liquids	Thermal aging of adsorbent Heat loss means inefficiency in energy usage Unsuitable for rapid cycling so adsorbent cannot be used with maximum efficiency In liquid systems high latent heat of interstitial liquid must be added.
Pressure swing	Good where weakly adsorbed species is required in high purity Rapid cycling-efficient use of adsorbent	Very low P may be required Mechanical energy more expensive than heat Desorbate recovered at low purity
Purge stripping	Operation at constant temperature and total pressure	Large purge volume required
Displacement	Good for strongly held species Avoids risk of cracking reactions during regeneration Avoids thermal aging of adsorbent	Product separation and recovery needed (choice of desorbent is crucial)

can reduce the original solution volume but requires an additional step to regenerate the second fluid (i.e., generally not a good idea).

Table 7.5 shows the advantages and disadvantages of the various regeneration modes.

7.7 Transport processes

Adsorption is typically operated as an equilibrium-limited process; the adsorbent must be in equilibrium with the surrounding fluid phase to obtain the maximum adsorption. It is also important to consider the various transport processes involved and the rate at which adsorption will occur. The mass transfer mechanism of adsorption typically has four steps (Figure 7.2).

1   Transfer of the solute (adsorbate) from the bulk fluid phase to the surface film (boundary layer) which surrounds the adsorbent particle. This step is controlled by convective flow and turbulent mixing.

2   Transfer of the adsorbate across the surface film to the exterior surface of the adsorbent particle. This step is controlled by molecular diffusion and/or convective flow.

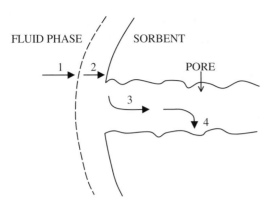

Figure 7.2 Adsorption mechanisms.

3 Transfer of the adsorbate from the particle surface to the interior of the adsorbent via the pore network. This step can be accomplished in two ways: pore diffusion (diffusion through the fluid in the pore); and surface diffusion (the particle travels along the pore surface).

4 Physical or chemical binding of the adsorbate to the internal surface of the adsorbent. This step is controlled by the molecular interactions described previously for adsorption.

Steps 1 and 4 are usually the fastest steps, and therefore are not considered to contribute to the overall rate of adsorption. The rate-determining step is typically Step 3, although changes in fluid flowrates can affect mass transfer across the fluid–particle boundary layer (Step 2).

For fixed beds, Step 2 can be described by:

$$j = 1.17Re^{-0.415} \quad 10 < Re < 2500 \tag{7.1}$$

$$= \frac{k}{v_{\text{s}}} Sc^{0.067} = \quad \text{Chilton–Colburn } j \text{ factor} \tag{7.2}$$

where $Re = \dfrac{\rho_{\text{f}} v_{\text{s}} d_{\text{p}}}{\mu}$ = Reynolds number

$\quad Sc = \dfrac{\mu}{\rho D_{AB}}$ = Schmidt number

$\quad \rho_{\text{f}}$ = fluid density

$\quad \mu$ = fluid viscosity

$\quad v_{\text{s}}$ = fluid superficial velocity

$\quad d_{\text{p}}$ = particle diameter

$\quad k$ = mass transfer coefficient

$\quad D_{AB}$ = diffusion coefficient of sorbate in fluid.

High values of k correspond to less mass transfer resistance for this step. One way to increase k is to increase v_{s}. This can cause problems, though, since the contact time in the bed would be reduced. This can lead to breakthrough with a larger portion of the bed unused.

A better alternative is a low value of v_s. This approach provides more contact time for sorbate–sorbent equilibrium and a lower pressure drop across the bed. In practice, this translates to the use of short bed lengths of large diameter (low v_s), in contrast to long, small-diameter beds (high v_s). The pressure drop across the bed can be estimated using the Ergun equation:

$$\frac{\Delta P}{L} = \left(150\frac{1-\varepsilon}{Re} + 1.75\right)\frac{\rho_f v_s^3(1-\varepsilon)}{d_p \varepsilon^3}, \tag{7.3}$$

where L = actual bed length

ε = bed void fraction (not particle).

Remember that the total pressure drop includes accounting for auxiliary components (valves, piping, etc.).

Step 3 depends on the sorbent size and the effective diffusivity (D_{eff}) within the sorbent particle, which can be written as

$$D_{eff} = \frac{D_{AB}\varepsilon_p}{\tau}, \tag{7.4}$$

where ε_p = particle void fraction

τ = tortuosity (correction factor > 1 to account for the tortuous nature of pore structure).

Appendix C describes pulse analysis that can be used to obtain process parameters. One measure of the ability of the sorbate to access the particle interior is D_{eff}/r where r is the characteristic sorbent size (radius for spheres). A large value of this parameter translates to good interior access. This result favors small particle size. This result is usually outweighed by pressure-drop considerations since a larger ΔP is needed as the particle size is reduced.

7.8   Process design factors

Before considering the specifics of adsorption design factors, it may be useful to generalize the process with some simplified analogies.

First, think of a large department store that has a parking lot next to it (Figure 7.3(a)). Before the store opens on a busy day, the parking lot is empty. When the store opens, cars arrive and typically park very close to the store. As more cars arrive, they have to park further and further away. The car traffic is a dynamic situation with cars coming (at an assumed constant rate) and going but one can observe that there is a net accumulation of cars as the lot fills up. At some time during this period, if we plot the number of cars vs position relative to the store (Figure 7.3(b)), the plot will look very similar to the one described for an adsorption column (see Figure 7.5). As the lot fills up, cars will enter and leave without stopping to park since the open spaces are isolated, far from the store

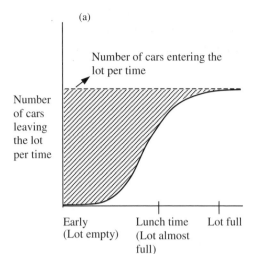
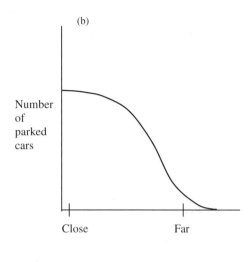

Figure 7.3 Parking lot analogy.

and may be difficult to access. So, Figure 7.3(a), the outflow will increase before the lot is completely full. The cross-hatched area in Figure 7.3(a) represents the total number of parked cars in the lot.

Next, imagine a beaker of water containing an organic pollutant to which a large amount of activated carbon is added. The pollutant will sorb onto, and desorb from, the carbon particles until the rate of sorption equals the rate of desorption and the system is in equilibrium. At equilibrium, the concentration of the pollutant in the water will be at a minimum value. The carbon in equilibrium with this solution will have reached its adsorptive capacity, and cannot adsorb any more pollutant under the current conditions. This batch system is one method to measure the equilibrium loading of the sorbent.

Now imagine a column filled with activated carbon through which the aqueous stream containing the organic pollutant enters at the top. The top layers of the carbon will initially remove the pollutant from the water, and will do so until that layer of carbon reaches its capacity. The next section of carbon will then begin to remove the pollutant. If the column is long enough, one can imagine a mass transfer zone, where all the adsorption is occurring. Below this zone, the contaminant in the water will have been reduced to its minimum value. Above the mass transfer zone, the carbon (which is saturated, i.e., at its adsorptive capacity) is in equilibrium with the influent pollutant concentration and mass transfer is no longer occurring. The actual height of the mass transfer zone will vary with flowrate. Typically, you want this zone as small as possible (can you see why?). In the limit of very low flowrates, the length of the mass transfer zone will reduce to zero and the concentration profile will propagate as a front. This position is called the stoichiometric front since it can be calculated from a mass balance (see Problem 7.1).

The mass transfer zone will move down the column with time. When it reaches the bottom of the column, the effluent pollutant concentration will start to rise above the minimum value, and this is known as *breakthrough*. When the very bottom layers of carbon reach their adsorptive capacity and the mass transfer zone disappears, the effluent pollutant concentration will reach the inlet concentration and the column is said to be *exhausted*.

Adsorbent columns can be used until they are exhausted by placing them in series. As one column is exhausted, it is taken off line and the adsorbent is replaced or regenerated. The column can then be placed below the second column, and the process is continued. If only one column were to be used, it would have to be regenerated sometime before exhaustion, depending on the allowable effluent pollutant concentration. Placing columns in parallel is also a possibility, so that breakthrough in one column will not significantly affect the effluent quality.

7.8.1   The mass transfer zone and breakthrough

We can use two methods to visualize the effect of the mass transfer zone (MTZ) on the adsorption column operation. In one method, we measure the solute concentration in the fluid phase at the column exit (position is fixed, time is a variable).

Figure 7.4 shows a typical result. The effluent solute concentration is at some minimum level (C_d) for a period of time until the concentration starts to rise as the MTZ arrives at the column exit. When the concentration rises to the maximum allowable effluent solute concentration (C_{bt}), breakthrough is said to occur. Column loading is defined as the amount of sorbed material in the bed at the breakthrough divided by the total weight of sorbent in the bed. If the bed were continued in operation past time t_{bt}, the effluent solute concentration would continue to rise until the entire bed was completely loaded and the effluent solute concentration equaled the feed solute concentration (C_f). This situation corresponds to column exhaustion. Note that an analogous plot could be made using total fluid volume processed instead of time as a variable.

The second method is to plot the solute concentration in the fluid phase as function of position (distance, L, from inlet) in the column for a fixed time that is less than t_d. Figure 7.5 shows a typical plot.

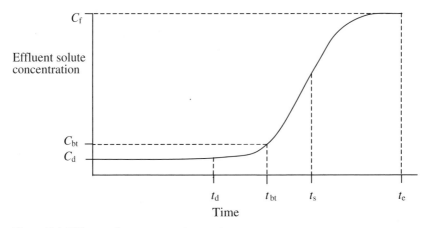

Figure 7.4 Effluent solute concentration vs time.

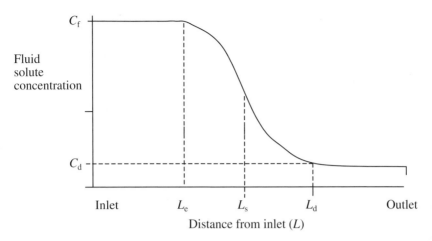

Figure 7.5 Fluid solute concentration vs position in the adsorption column.

From the column inlet to L_e, the sorbent is loaded to capacity and the solute concentration in the fluid phase is C_f. From L_e to L_d, the MTZ exists where adsorption is occurring. From L_d to the column exit, the fluid solute concentration is C_d. Refer back to the parking lot analogy with discussion in this section (compare Figure 7.3(a) and (b) with Figures 7.4 and 7.5, respectively).

The various transport processes described in Section 7.7 affect the length of the MTZ. As the fluid dispersion and/or the diffusional mass transfer resistances increase, the length of the MTZ will increase. We will see how the adsorption isotherm affects the MTZ in Section 7.9.1. Also, in Section 7.10, we will see how the breakthrough curves (Figure 7.4) can be used for column design and scale-up. It is important to note that if the flow conditions are changed (particle size, sorbent, etc.) then the mass transfer characteristics of the column will change and, consequently, the resulting breakthrough curve will also change (i.e., be very careful when using these methods).

If the fluid solute concentration in the column propagates as a stoichiometric front (plug-flow), the position of the front would be L_s in Figure 7.5. This corresponds to the portion of the column from L_s to L_d as being unused. When L_d corresponds to the column length L_c, the length of the unused bed (LUB) can be defined as

$$LUB = \left[1 - \frac{L_s}{L_c}\right] L_c. \tag{7.5}$$

Defining t_s as the time that the stoichiometric front would arrive at the column exit, an analogous expression using the plot in Figure 7.4 is

$$LUB = \left[1 - \frac{t_d}{t_s}\right] L_c. \tag{7.6}$$

7.8.2 Pressure and temperature considerations

The temperature and pressure conditions are very important for column operation. If there is a temperature choice within some range, the coldest temperature is usually selected to obtain the highest sorption loading. The temperature of the fluid exit stream is the usual point for evaluation. The sorption process is exothermic so that the exit temperature is the highest in the system. One important exception is when the feed stream contains a condensible vapor that is not to be separated out of the feed stream. The system temperature must be maintained <u>high</u> enough so that condensation does not occur. If the condensible vapors are to be separated, it is usually easier to do so by condensation (using an energy-separating agent) prior to the sorption step.

For pressure considerations, the highest pressure is at the feed entrance since the pressure will decrease due to flow through the system. For a system that contains condensible vapors, the pressure must be maintained <u>low</u> enough so that condensation does not occur.

7.8.3 System orientation and flow direction

The system should normally be oriented in the vertical direction. The sorbent particles will settle in the vessel with time and use. If the system were oriented horizontally, some open space would develop at the top of the vessel and the fluid would tend to flow through this section due to decreased flow resistance (channeling). This channeling effect is usually very detrimental to separation performance and cycle lifetime.

The second factor to consider is the direction of the fluid flow, upward or downward. Upward flow will cause lift of the particle bed at some critical velocity that causes the bed to fluidize. This effect results in an increase in the fluid dispersion within the system and a decrease in performance. Downward flow could cause the particles to be crushed at some point. The allowable velocities for crushing are larger than those for lift, so downward flow is the normal operating condition.

7.8.4 Flow dispersion

Axial dispersion (sometimes referred to as backmixing) is a spreading of the concentration profile in the axial direction due to flow variations within the adsorbent bed (see the pulse analysis section in Appendix C). This effect can also contribute to the spreading of the mass transfer zone.

7.8.5 Non-isothermal effects

Many adsorption design approaches assume that adsorption is occurring isothermally. This is a good assumption only when the adsorbable component concentration is low and/or the heat of adsorption is low. There are two simple methods that can be used to determine if the system can be assumed to be isothermal. Since most of the heat is generated in the MTZ where adsorption is occurring, the rate at which the heat can be transferred out of this zone compared to the movement of the MTZ is the basis for one method [7]. This comparison is shown by the "crossover ratio" R:

$$R = \frac{C_{pf}(X_i - X_{res})}{C_{ps}(Y_i - Y_o)} \tag{7.7}$$

where Y is the molar ratio of sorbate to the carrier fluid (i denotes inlet, o denotes outlet), and the fluid and sorbent heat capacities, C_{pf} and C_{ps}, include the effect of the sorbate. X is the sorbent loading (wt sorbate/wt sorbent); X_{res} is the residual loading in the bed prior to the adsorption step. When $R \gg 1$, the heat is easily removed from the MTZ and adsorption can be assumed to be isothermal. As R approaches a value of 1, more and more heat will be retained in the MTZ. An increase in the temperature of the "leading" or breakthrough end of the MTZ will lower the equilibrium loading from the isothermal value based on the inlet temperature and cause the curve to become less favorable relative to the operating line, until ultimately the MTZ has no stable limit but continues to expand as it moves through the bed. When $R = 1$, the heat front is moving through the bed at the same velocity as that of the MTZ, and essentially all the heat of adsorption is found in the MTZ. For cases where $R < 1$, the heat front will lag the adsorption front and heat will be stored in the equilibrium section. Here the temperature rise will cause the equilibrium loading to decrease. Thus, the crossover ratio is an indication of non-isothermal operation, the extent of the harmful effects of the temperature rise due to adsorption, and the location of the temperature change.

A second method computes the temperature rise under equilibrium conditions [10]

$$\Delta T = T_{max} - T_f = \frac{q\Delta H/C_{pg}}{(q/Y)_f - C_{ps}/C_{pg}} \tag{7.8}$$

where q = solute adsorbed/mass of sorbent

ΔH = heat of sorption

Y = mass solute in fluid phase/mass of carrier gas

C_p = heat capacity,

and the subscripts are:

s = sorbent

g = gas

f = feed.

For many operating conditions $C_{ps}/C_{pg} \ll (q/Y)_f$, so the above equation reduces to

$$\Delta T = Y_f \Delta H/C_{pg}. \tag{7.9}$$

A quick estimate of ΔT for gas-phase sorption can be made. ΔH has a range of 1000 to 4000 kJ/kg (avg = 2500), C_{pg} is approx. 1 kJ/kg · K and Y_f is typically 0.01. Therefore,

$$\Delta T = \frac{\left(0.01\,\frac{\text{kg solute}}{\text{kg carrier gas}}\right)\left(2500\,\frac{\text{kJ}}{\text{kg solute}}\right)}{1\,\frac{\text{kJ}}{\text{kg carrier gas·K}}} = 25\,\text{K}\,(\equiv\,25\,°\text{C}).$$

So the maximum temperature rise in a gas–solid sorption can reach 25 °C.

A few additional points are worth noting:

1 For constant partial pressure, Y_f is inversely proportional to total pressure. Therefore, ΔT will decrease with an increase in total pressure. This translates to isothermal operation for a total feed pressure of approximately 50 atm.

2 Carrier gases with high C_{pg} (hydrogen, for example) will tend to reduce ΔT.

3 An increase in feed solute concentration will increase ΔT.

7.8.6 Bed stability

Not all adsorption beds will develop stable MTZs. One requirement for stability (i.e., the MTZ reaches a limiting size) is that the equilibrium line must be "favorable." In the case of a single adsorbate isothermally removed from a non-adsorbable component, the curve of loading as a function of composition must be concave downward in the region of loading below the stoichiometric point to be favorable. This effect is described in more detail in Section 7.9. In non-isothermal adsorption it is possible for the temperature effects to cause a favorable isotherm to become an unfavorable equilibrium line. This was discussed previously in the context of the crossover ratio R.

7.8.7 Special considerations for liquids

Flow direction considerations for liquid systems are somewhat different than those for vapor flow. In liquid or dense-phase flow the buoyancy force of the liquid must be considered as well as the pressure drop. During upflow adsorption, the flow velocity should be low so as to not cause bed expansion (fluidization). As the flowrate exceeds this limit, the pressure drop increase is small with increasing velocity. Sometimes liquid systems are designed with some bed expansion (10% at the most) when it is desirable to limit pressure drop. Upflow is preferred if the liquid contains any suspended solids, so that the bed will not become plugged.

Prior to the introduction of a liquid into the adsorbent bed, there must be sufficient time for any gas or vapor that may be trapped in the pores of the sorbent to outgas. Otherwise the gas or vapor may contaminate a product during operation. In the case of upflow adsorption, the effective bulk density may be lowered enough to cause excessive bed expansion or flow channeling.

Upon completion of the adsorption step, draining of the adsorption liquid is usually done by gravity flow, sometimes assisted by a pressure of 10–20 psig (69–138 kPa). The liquid must be given at least 30 minutes to drain thoroughly. Even then, there can be significant hold-up or retention of liquid in the bed. Even after careful draining, hold-up can amount to 40 cm^3 of fluid per 100 g of adsorbent. This fluid is retained in the micro- and macropores and bridges between particles. Retained liquids can adversely affect product streams and regeneration requirements. For example, any liquid that is not drained in a temperature-swing cycle will consume extra thermal energy when it is vaporized from the bed, and the fluid will end up recovered with the adsorbate.

In liquid adsorption systems where liquids are also used for purge or displacement, care must be taken to prevent "fingering." Fingering is the displacing of one liquid by another at their interface due to density or viscosity differences. The phenomenon creates columns of the intruding fluid (fingers) even in uniformly packed beds of adsorbent. It is obvious that a denser fluid above a less dense fluid will cause instability. However, it is also true that when a less viscous fluid is displacing a more viscous one, any bulge in the interface will grow because the resistance to flow is less, and the less viscous fluid will continue to intrude. Operating such that the upper fluid is the less dense or more viscous fluid for displacing, will tend to correct any flow instabilities that occur.

7.9 Evaluating the adsorption process

An evaluation of the adsorption (and desorption) steps can be accomplished using the equilibrium isotherms. This will be discussed in this section. Two simplified analyses (scale-up and kinetic) based on obtaining a breakthrough curve in a small, laboratory-scale apparatus and using the results for the design of larger, process-scale unit will be discussed in Section 7.10. A detailed approach that considers the various transport steps allows performance to be predicted by the model under varying flow conditions. The reader should consult Yang's text [4] for a detailed presentation.

7.9.1 Equilibrium-limited operating conditions

We can use the equilibrium isotherm to predict performance under equilibrium-limited operating conditions [10]. A mass balance on the solute in the MTZ leads to an equation for the velocity of the front:

Amount introduced = Amount retained

$$YG_bAt = q\rho_sAz \tag{7.10}$$

$$V = \frac{dz}{dt} = \text{velocity of solute front} = \frac{G_b}{\rho_s \cdot dq/dY} = \frac{\rho_f v}{\rho_s \cdot dq/dY}, \tag{7.11}$$

where Y = mass of solute/mass of fluid

ρ_f = fluid density

v = fluid velocity

q = mass sorbed/mass of sorbent

ρ_s = mass of sorbent/volume

G_b = bulk fluid mass flux = mass of fluid/area · time

L = bed length (see Equation (7.12))

A = column cross-sectional area.

This result indicates that the velocity of the solute front is inversely proportional to the slope of the isotherm. We can illustrate this result using a Type I isotherm (Figure 7.6). During the adsorption step, the direction is from the lower left to upper right portion of the curve. So, dq/dY is largest (V is slowest) during the initial portion of sorption. This is the rate-limiting step so the entire front moves as a discontinuous wave (stoichiometric front). A balance across this wave shows that dq/dY reduces to $\Delta q/\Delta Y$, the chord from the initial state to the saturated state in the column.

In desorption, the result is very different. Now, the direction is from the upper right to the lower left portion of the curve. The slope dq/dY has the smallest value (V is fastest) during the initial portion and V becomes slower as desorption continues. This causes the desorption wave to spread and an elongated breakthrough curve results.

Some additional system parameters can also be estimated (Equations (7.12) to (7.16)). The minimum weight of sorbent needed to treat a given total weight of fluid is $\Delta Y/\Delta q$,

$$\frac{A\rho_b L}{A\rho_f vt} = \frac{\text{bed weight}}{\text{weight of carrier fluid processed}} = \frac{\Delta Y}{\Delta q} = \frac{Y_f - Y_i}{q_f - q_i}, \tag{7.12}$$

where Δ is the difference between the feed (f) and initial (i) column conditions.

For desorption, first recognize that $dq/dY = H$ (Henry's Law constant) as the origin of the isotherm is approached. So, the minimum amount of purge fluid needed per weight of column sorbent equals H. The minimum time for breakthrough can then also be determined:

$$\frac{dq}{dY} = H (\text{as you approach origin} \rightarrow \text{for total desorption}). \tag{7.13}$$

$$\therefore \frac{1}{H} = \frac{A\rho_b L}{A\rho_f vt} \tag{7.14}$$

$$H = \frac{\text{weight of purge required}}{\text{bed weight}} = \frac{\rho_f vt}{\rho_b L} = \frac{G_b t}{\rho_b L}. \tag{7.15}$$

Rearranging,

$$t = \frac{\rho_b H}{G_b} L. \tag{7.16}$$

Results for various isotherms are shown in Figures 7.7 to 7.10.

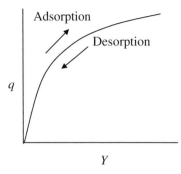

Figure 7.6 Type I isotherm to demonstrate movement of solute front.

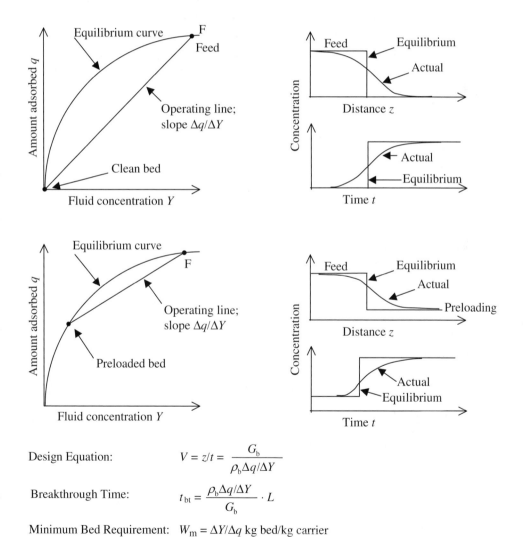

Design Equation:

$$V = z/t = \frac{G_b}{\rho_b \Delta q/\Delta Y}$$

Breakthrough Time:

$$t_{bt} = \frac{\rho_b \Delta q/\Delta Y}{G_b} \cdot L$$

Minimum Bed Requirement: $W_m = \Delta Y/\Delta q$ kg bed/kg carrier

Figure 7.7 Type I isotherm: adsorption [10].

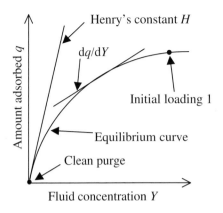

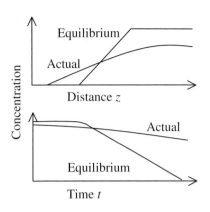

Design Equation:
$$V = z/t = \frac{G_b}{\rho_b dq/dY}$$

Desorption Time:
$$t_{dt} = \frac{\rho_b H}{G_b} \cdot L$$

Minimum Bed Requirement: $W_m = 1/H$ kg bed/kg carrier

Figure 7.8 Type I isotherm: desorption [10].

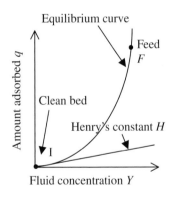

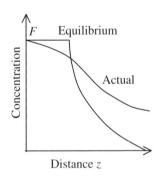

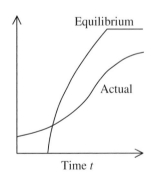

Design Equation:
$$V = z/t = \frac{G_b}{\rho_b dq/dY}$$

Breakthrough Time:
$$t_{bt} = \frac{\rho_b H}{G_b} \cdot L$$

Minimum Bed Requirement: $W_m = 1/H$ kg bed/kg carrier

Figure 7.9 Type III isotherm: adsorption [10].

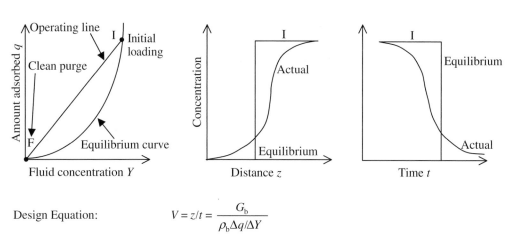

Design Equation:
$$V = z/t = \frac{G_b}{\rho_b \Delta q/\Delta Y}$$

Desorption Time:
$$t_{dt} = \frac{\rho_b \Delta q/\Delta Y}{G_b} \cdot L$$

Minimum Purge Requirement: $G_m = \Delta q/\Delta Y$ kg purge/kg bed

Figure 7.10 Type III isotherm: desorption [10].

7.10   Design of fixed-bed adsorption columns

Two macroscopic methods to design adsorption columns are the scale-up and kinetic approaches. Both methods rely on breakthrough data obtained from pilot columns. The scale-up method is very simple, but the kinetic method takes into account the rate of adsorption (determined by the kinetics of surface diffusion to the inside of the adsorbent pore). The scale-up approach is useful for determining the breakthrough time and volume (time elapsed and volume treated before the maximum allowable effluent concentration is achieved) of an existing column, while the kinetic approach will determine the size requirements of a column based on a known breakthrough volume.

7.10.1   Scale-up approach

Initially, a pilot column with a bed volume (V_p) and volumetric flowrate of fluid (Q_p) is used. As shown in Figure 7.11, the total volume (V_T^{pilot}) of fluid that passes through the column is measured until the outlet solute concentration is observed to rise to the maximum allowable value (C_a).

The (plant-scale) design column should operate such that:

1 Fluid residence time in the pilot and design column are the same.
2 The total volume of fluid processed until breakthrough per mass of sorbent in the column is the same for both columns.

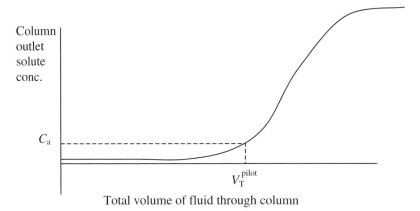

Figure 7.11 Outlet concentration vs volume throughput for pilot column.

The parameters needed to effect the scale-up are:

$(BV) = (V_p)Q/Q_p = $ Bed volume of design column, where Q is the fluid volumetric flowrate of the design column

$M = (BV)(\rho_s) = $ mass of adsorbent in the design column ($\rho_s = $ adsorbent bulk density)

$V_T^{pilot} = $ breakthrough volume of the *pilot* column (chosen to correspond to the maximum allowable effluent solute concentration). This volume is the total amount treated before breakthrough occurs in the pilot column.

$\hat{V}_B = \dfrac{V_T^{pilot}}{M^{pilot}} = $ volume of liquid treated per unit mass of adsorbent (same for both columns)

$M_t = Q/\hat{V}_T = $ mass of adsorbent exhausted per hour in design column Q.

$t_{bt} = M/M_t = $ breakthrough time for the design column

$V_T = Qt_{bt} = $ breakthrough volume for the design column.

Example 7.1: Fixed-bed column design by the scale-up approach

Problem:
A wastewater flowrate of 180 m³/day has a TOC (total organic carbon level) of 200 mg/L. A fixed-bed GAC adsorption column will be used to reduce the maximum effluent concentration to 8 mg/L. A breakthrough curve, Figure 7.12, has been obtained from an experimental pilot column operated at 2(BV)/hr. Other data concerning the pilot column are: mass of carbon = 4.13 kg, water flowrate = 15 L/hr, and packed carbon density = 400 kg/m³. Using the scale-up approach, determine the values of the following parameters for the design column:
(a) Bed volume (BV)
(b) Mass of carbon required (M)

(c) Breakthrough time (t_{bt})

(d) Breakthrough volume (V_T).

Solution:

(a) $(BV) = \dfrac{Q}{Q_b} = \dfrac{180\,\text{m}^3}{24\,\text{hrs}}\dfrac{\text{hr}}{2.0(BV)} = 3.75\,\text{m}^3$

(b) $M = (BV)(\rho_s) = (3.75\,\text{m}^3)(400\,\text{kg/m}^3) = 1500\,\text{kg}$

Since $V_T^{\text{pilot}} = 2000\,\text{L}$ at $C = 8\,\text{mg/L}$, $\dfrac{V_B^{\text{pilot}}}{M^{\text{pilot}}} = \dfrac{2000\,\text{L}}{4.13\,\text{kg}} = 484\,\text{L/kg}$

$M_t = \dfrac{Q}{\hat{V}} = \dfrac{180\,\text{m}^3}{24\,\text{hrs}} \cdot \dfrac{1\,\text{kg}}{484\,\text{L}} \cdot \dfrac{1000\,\text{L}}{\text{m}^3} = 15.5\,\text{kg/hr}$

(c) $t_{bt} = \dfrac{M}{M_t} = (1500\,\text{kg})\left(\dfrac{\text{hr}}{15\,\text{kg}}\right) = 100\,\text{hrs or approx. 4 days}$

(d) $V_T = Qt_{bt} = \left(\dfrac{180\,\text{m}^3}{24\,\text{hrs}}\right)(100\,\text{hrs}) = 750\,\text{m}^3$.

<u>Note</u>: If you are given the cross-sectional area of the pilot column, you can calculate the diameter of the design column. The volumetric flowrate of each column is known and the velocity in each column is the same. The design column length can then be calculated since the bed volume is known.

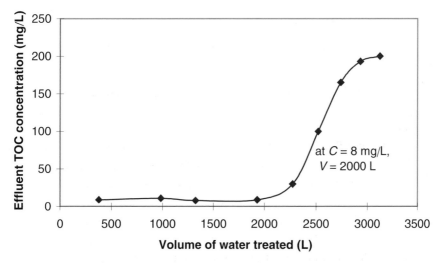

Figure 7.12 Breakthrough curve for TOC removal from wastewater by GAC adsorbent (Example 7.1).

Table 7.6 *Data from pilot column breakthrough test, Example 7.2.*

Volume of water treated, V (L)	Effluent solute concentration, C (mg/L)
1500	0.3
1750	2.4
1900	7.9
2200	65.5
2400	143.0
2500	170.1
2600	185.6
2800	197.0
3000	199.4
3200	199.9
3300	200.0

7.10.2 Kinetic approach

If the design fluid volumetric flowrate (Q) is sufficiently low that equilibrium is rapid in comparison, the Equation (7.17) below is a good approximation of the concentration profile for the breakthrough curve as a function of fluid volume (V) put through the column [11]. A Langmuir isotherm is assumed where k_1 is the adsorption rate constant for this isotherm. When $q_0 M \gg C_0 V$, the effluent solute concentration is approximately zero. For $q_0 M \ll C_0 V$, the effluent solute concentration is C. See for yourself why this makes sense physically.

$$\frac{C}{C_0} \cong \frac{1}{1 + \exp\left[\dfrac{k_1}{Q}(q_0 M - C_0 V)\right]},$$

(7.17)

where C = effluent solute concentration

$\qquad C_0$ = influent solute concentration

$\qquad k_1$ = rate constant

$\qquad q_0$ = maximum solid-phase concentration of the sorbed solute (g/g or lb/lb)

$\qquad M$ = mass of the adsorbent

$\qquad V$ = throughput volume

$\qquad Q$ = fluid volumetric flowrate.

This equation can be rewritten as:

$$\ln\left(\frac{C_0}{C} - 1\right) = \frac{k_1 q_0 M}{Q} - \frac{k_1 C_0 V}{Q},$$

(7.18)

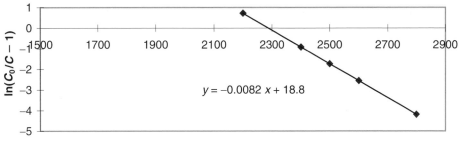

Figure 7.13 Breakthrough curve data applied to the estimation of the rate constant k_1, Example 7.2.

which is a straight line on a plot of effluent solute concentration vs volume of fluid treated, and the pilot column breakthrough data can be used to determine k_1, q_0, and other parameters of the design column.

Example 7.2: fixed-bed column design by the kinetic approach

Problem:

Redo the previous example using the kinetic approach. Given the data in Table 7.6, determine the following values:

(a) k_1(reaction constant); (b) q_0; and (c) mass of carbon required in the design column

Solution:

The points which pertain to the breakthrough portion of the data are plotted in the appropriate form in Figure 7.13. Note that it is a straight line.

(a) Remember from the previous Equation (7.17) that:

$$\text{Slope} = \frac{-k_1 C_0}{Q},\tag{7.19}$$

or

$$k_1 = 0.0082 \text{ L}^{-1}(15 \text{ L/hr}) (\text{L}/200 \text{ mg}) = 6.15 \times 10^{-4} \text{ L/hr} \cdot \text{mg}.$$

(b) Remember also that

$$\text{Intercept} = \frac{k_1 q_0 M}{Q},\tag{7.20}$$

or

$$q_0 = (18.8)(15 \text{ L/hr})(\text{hr} \cdot \text{mg}/6.15 \times 10^{-4} \text{ L})(1/4.13 \text{ kg})$$
$$= 1.11 \times 10^5 \text{ mg TOC/kg GAC}.$$

(c) Substituting the design volume into Equation (7.18) and solving for M:

$$Q = (2(BV)/\text{hr})(3.75 \text{ m}^3) \times 1000 \text{ L/m}^3 = 7.5 \times 10^3 \text{ L/hr}$$

$$\ln\left(\frac{200}{8} - 1\right) = \left(6.15 \times 10^{-4} \frac{L}{hr \cdot mg}\right)$$

$$\times \left(1.11 \times 10^5 \frac{mg\ TOC}{kg\ GAC}\right) M / 7.5 \times 10^3\ L/hr$$

$$- \frac{\left(6.15 \times 10^4 \frac{L}{hr \cdot mg}\right)\left(200 \frac{mg\ TOC}{L}\right)(720\ m^3)\left(\frac{1000\ L}{m^3}\right)}{7.5 \times 10^3\ L/hr}.$$

Therefore, $M = 1.65 \times 10^3$ kg, close to the value calculated in Example 7.1.

7.11 Remember

- Adsorption is a UNIT OPERATION. Regardless of what chemicals are being separated, the basic design principles for adsorption are always similar.
- Adsorption requires an interface between phases (solvent and adsorbent), and a driving force must exist for the adsorbate to accumulate on the adsorbent. Electrostatic forces or chemical bonding reactions are examples of the necessary driving force which allows adsorption to occur.
- Adsorption can be reversible or irreversible; it is sometimes possible to regenerate the adsorbent (usually some losses in the amount of adsorbent and its capacity to adsorb are encountered).
- Adsorption has been used in wastewater treatment primarily for taste and odor control, but it is growing more popular for removal of contaminants such as synthetic organic chemicals, color-forming organics, disinfection chemicals and their by-products (the most notorious being the trihalomethanes), and heavy metals.
- Adsorption may be chosen over distillation when: (1) undesirable reactions occur during distillation; (2) an azeotrope is encountered or when the boiling range of one set of components overlaps the range of another set; (3) throughputs are less than a few tons per day; or (4) corrosion, precipitation, or explosive conditions make distillation impossible.
- An adsorption isotherm relates the amount of the substance adsorbed at thermodynamic equilibrium to the amount present in the liquid or gas stream (concentration or partial pressure) at a constant temperature. Langmuir and Freundlich isotherms are the most common adsorption isotherms.
- An adsorption isotherm is useful for scaling up small-scale batch processes usually carried out in a laboratory. Once the laboratory data are fitted to an isotherm, one can predict the amount of adsorbent required to reach a specific effluent solute concentration (in terms of a batch reactor) or the breakthrough time (for a plug-flow column).

7.12 Questions

7.1 Why would you want the mass transfer zone in an adsorption column to be as small as possible?

7.2 What are the differences between breakthrough and exhaustion?

7.3 What are the important criteria for choosing an adsorbent?

7.4 How would a lower amount of dispersion in an adsorption column affect the breakthrough curve?

7.5 Show that the limits described for Equation (7.17) are valid.

7.13 Problems

7.1 Show that a mass transfer zone that propagates as a stoichiometric front would correspond to the shortest sorbent bed needed.

7.2 Chlorides are removed from water with a carbon adsorbent. The carbon particle diameter is 0.2 cm, the viscosity and density of water are 0.8 cP and 1 g/cm^3, respectively, and the diffusion coefficient of the chlorides in water is 2.37×10^4 cm^2/s. Calculate the mass transfer coefficient and bed diameter to treat 125,000 cm^3/s water for superficial velocities of 5, 10, 25, 100, 250 cm/s. Explain the disadvantage of increasing the superficial velocity.

7.3 For Problem 7.2, estimate the pressure drop for $v_s = 10$ cm/s and 25 cm/s. By what factor does the pressure drop increase? Assume $L = 1$ m and bed void fraction $(\epsilon) = 0.5$.

7.4 Aqueous effluent from a processing plant contains 300 mg/L toluene that is to be reduced to 15 mg/L prior to discharge. 640,000 L of a 400 L/min stream need to be treated prior to breakthrough. The rate constant, k_1, is 5.56 L/kg · min and q_0 is 0.2 g/g based on small-scale studies. What mass of adsorbent is required in the design adsorption column?

8

Ion exchange

Discussion is an exchange of knowledge; argument an exchange of ignorance.

– ROBERT QUILLEN

8.1 Objectives

1 Describe the mechanism of ion exchange.
2 List the types of solutions which ion exchange is capable of separating.
3 Differentiate between:
   (a) Strong-acid cation exchangers;
   (b) Weak-acid cation exchangers;
   (c) Strong-base anion exchangers;
   (d) Weak-base anion exchangers.
4 Compare ion exchange and adsorption.
5 Design an ion-exchange column via the same kinetic approach used in the design of adsorption columns.

8.2 Background

Ion exchange is very similar to adsorption; both processes involve mass transfer from a fluid to a solid phase. Ion exchange can be described as a sorption process, but ions are sorbed in comparison to electrically neutral species in adsorption. An important difference between ion exchange and adsorption is that ion exchange requires that the species removed from the fluid phase is replaced with a species (exchanged) so that electroneutrality is maintained. Electroneutrality requires that the total charge sorbed and desorbed is the same. For example, two Na^+ will exchange with one Ca^{2+}.

Ion exchange is becoming used more extensively in water and wastewater treatment. Ion exchange is primarily used for water softening (Ca^{2+} and Mg^{2+}) and for water demineralization. For water softening, Ca^{2+} and Mg^{2+} are replaced with Na^+ to prevent scale formation. For complete water demineralization, all cations and anions are replaced with H^+ and OH^-, respectively. This approach is also used in wastewater treatment. It is important to note that not all dissolved ions are removed equally and/or completely. Ions that are low on the selectivity preference order (described in Section 8.5) may not be completely removed.

8.3   Environmental applications

The treatment of mine drainage water, removal of ammonia and nitrates from groundwater, and the treatment of nuclear waste solutions are some examples of environmental applications. Ion-exchange resins can also be used for pollutant removal from gas streams. For example, H_2S and NH_3 have been removed using macroreticular carboxylic acid resins and quaternary ammonium anion-exchange resins, respectively. The selective removal of these two impurities in hydrogen-cycle gas streams from oil refinery processes and their subsequent recovery by thermal elution using an inert gas are one important application [1]. For these applications, additional processing steps are usually required to obtain the material in pure form or reduce the volume prior to disposal (water removal, ion separation, precipitation, etc.).

The efficiency of waste treatment is strongly dependent on the regenerant consumption. Success is likely if the process fluid phase is either acidic or basic since this will affect the initial ion exchange. Actual process design depends on the waste to be treated, pollutant concentration, flowrate, and other operating conditions.

8.4   Ion-exchange mechanisms

All ion exchangers, whether natural or synthetic, have fixed ionic groups that are balanced by counterions to maintain electroneutrality. The counterions exchange with ions in solution. As an example, consider the schematic cation-exchange resin shown in Figure 8.1.

The resin containing cation B^+ is placed in a solution containing cation A^+. The cations A^+ and B^+ will diffuse due to a concentration gradient between the resin and solution. The chemical equation for this particular exchange reaction within the ion-exchange resin is:

$$A^+ + (R^-)B^+ \leftrightarrow B^+ + (R^-)A^+, \tag{8.1}$$

where R^- represents the negatively charged functional group of the resin.

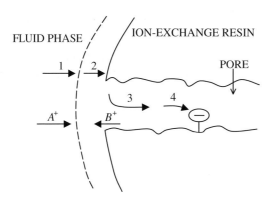

Figure 8.1 Schematic diagram of a cation-exchange resin.

An alternative equation is:

$$\bar{B}^+ + A^+ \leftrightarrow \bar{A}^+ + B^+. \tag{8.2}$$

Ion exchange will continue until the equilibrium described by Equation (8.1) is reached. Note that equilibrium does not imply equal concentrations of each ion in the resin and fluid phase. Also, ion diffusion is coupled with charge neutrality and not solely due to concentration differences.

Like adsorption, the mass-separating agent is the resin material. The transport steps that take place during ion exchange are also similar to adsorption.

1 Transport of the exchanging ions to and from the bulk solution to the surface film (boundary layer) surrounding the resin;
2 transport of the exchanging ions through the surface film (or boundary layer) at the external surface of the particle;
3 interstitial (pore) transport of the exchanging ions to the sites of active exchange; and
4 kinetics of the exchange process.

Again, as with adsorption, Steps 2 and 3 are typically the slowest and rate controlling. The nature of the rate-determining step can be predicted by use of the simple dimensionless criterion given by Helfferich [2, 3]:

$$\frac{\bar{C}\bar{D}\delta}{CDr_0}(5 + 2\alpha_{AB}) \ll 1 \quad \text{pore transport} \tag{8.3}$$

$$\frac{\bar{C}\bar{D}\delta}{CDr_0}(5 + 2\alpha_{AB}) \gg 1 \quad \text{boundary layer}, \tag{8.4}$$

where α is the separation factor, r_0 is the radius of an ion-exchange bead, δ is the boundary-layer thickness of the fluid adjacent to the resin particle surface, $\bar{C}$ is the resin-phase concentration of ions, $\bar{D}$ is the diffusion coefficient in the resin phase, C is the concentration of ions in the solution phase, and D is the diffusion coefficient in the solution phase. If film diffusion is much faster than diffusion within the ion-exchange particles, then concentration differences in the liquid are very small.

Concentration gradients can exist within the resin pore structure. Ion diffusion is complex since the resin porosity is low and this leads to steric hindrance effects and tortuous diffusion paths. Also, ion diffusion is coupled to the fixed ionic groups and the mobility of each ion within the resin due to charge balance. This coupled diffusion is present in both boundary layer and pore transport. Also, the forward and reverse rates of ion exchange can be affected by the different mobilities of the ions.

8.5 Ion-exchange media

Some ion-exchange resins occur naturally and have been used for hundreds of years. They include clay peat, charred bone, and natural aluminosilicates. The recognition of ion exchange as a process is generally attributed to H. S. Thompson and J. Thomas Way, who were English agricultural chemists. Thompson observed in 1848 that soil treated with either ammonium sulfate or ammonium carbonate adsorbed the ammonia and released lime. He reported his results to Way and then conducted systematic studies, 1850–54. In 1935, B. A. Adams and E. L. Holmes observed that crushed phenolic phonograph records were capable of ion exchange. This observation led to the development of synthetic organic ion-exchange resins. With this development, the industrial use of ion exchange was rapidly increased.

Resins are typically synthesized by copolymerization of styrene and divinylbenzene (DVB). The styrene provides the backbone, and DVB is used as a crosslinker to stabilize the structure. The resin is then reacted with an acid or base to produce the fixed charged groups. Crosslinking varies radially within the resin. The extent is usually described by a 'nominal DVB content'. The degree of crosslinking is important because it determines the internal pore structure (see transport Step 3 above). The greater the percentage of DVB, the less the resin will swell when ions are exchanged, but the resin will have a tight pore structure with low mass transfer rates. Commercial resins are 2 to 12% DVB.

Resin beads are synthesized as gel or macroporous materials. The macroporous resins are polymerized in the presence of a third component that is insoluble in the polymer. After this insoluble component is removed, large pores remain that allow the ions to have improved access to the interior pore structure of the beads. Macroporous resins can be useful for large ions like proteins, but they are more expensive, have lower capacity, and are harder to regenerate than the gel resins. However, they are said to be more resistant to thermal and osmotic shock as well as to oxidation and organic fouling than the gel-type resins [4].

The following factors are important in the choice of an ion-exchange resin:
1 Exchange capacity (loading or productivity).
2 Fraction or percent removal of various ions from the liquid phase (selectivity).
3 Particle size and size distribution (flow throughput considerations).
4 Chemical and physical stability.
5 Regeneration requirements (chemicals, amounts required, loss in capacity).

Capacity is the quantity of the counterion that the resin can exchange. It is a critical factor in evaluating a resin for a given application. The total capacity is determined by the number and charge on the fixed ion-exchange groups in the resin. The dry-weight capacity is then determined as the milliequivalents per gram of dry resin (meq/g). This quantity is a measure of the loading capacity of the resin and is a constant.

The wet-exchange capacity accounts for the fact that the resin will swell or shrink during operation. This quantity will vary with moisture content. It is usually reported as equivalents per liter of resin (eq/L). Units of kilograms of $CaCO_3$ per cubic foot (kg/ft^3) are also used. It is this term that is typically reported.

Selectivity is primarily dictated by ionic charge and size, with charge having the most significant effect. Kunin [5] proposed the following empirical points to approximate selectivities.

1 At low aqueous concentrations and ambient temperatures, the extent of exchange increases with increasing valence of the exchanging ion:

$$Th^{4+} > Al^{3+} > Ca^{2+} > Na^+$$
$$PO_4^{3-} > SO_4^{2-} > Cl^-.$$

2 At low aqueous concentrations, ambient temperatures and constant valence, the extent of exchange increases with increasing atomic number (decreasing hydrated radius) of the exchanging ion:

$$Cs^+ > Rb^+ > K^+ > Na^+ > Li^+$$
$$Ba^{2+} > Sr^{2+} > Ca^{2+} > Mg^{2+} > Be^{2+}.$$

3 At high ionic concentrations, the difference in exchange "potentials" of ions of different valence (Na^+ vs Ca^{2+} or NO_3^- vs SO_4^{2-}) diminish and, in some cases, the ion of lower valence has the higher exchange potential.

It is important to note that these are "rules-of-thumb" and exceptions do occur.

Another aspect of selectivity is ion exclusion. Large organic ions or inorganic complexes can be excluded from the resin pore structure. Obviously, the smaller the pore size, the larger the potential for this effect.

Particle size affects the pressure drop through the column. Smaller particle size leads to higher pressure drops for a given flowrate. Often, hydraulic limitations are the most important consideration in design. Particle size also affects the relative magnitude of transport Steps 2 and 3 listed above. This is analogous to the discussion on this point in Chapter 7: Adsorption.

Stability is directly related to the lifetime of the resin. This, in turn, directly affects the cost of the process. Physical stresses can occur through swelling and shrinking cycles due to osmotic pressure changes. Mechanical forces, such as static pressure load, and abrasion can cause breakage. Operation outside the normal temperature range will also add to particle degradation.

Fouling, modification of the ion-exchange site or breakage of the resin structure can be caused by chemical degradation processes. Organic acids are charged and can exchange irreversibly onto strong-base resins. Also, silica fouling can occur in strong-base (OH^-) resins. The acidic silica concentrates at the ion-exchange front and forms a solid. Precipitates can also form in the resin that will restrict or block pores. Oxidative processes can attack the resin structure. This typically reduces the rigidity of the structure and increases swelling. The functional groups can undergo modification over time. This modification is accelerated as the operating temperature is increased.

Regeneration typically involves several steps. At the end of the exchange cycle, a backwash is typically done. This step serves two purposes. First, trapped particles in the bed are removed. Second, the backwash serves to remix the bed and reclassify the particles so that there is a gradual increase in particle size from top to bottom (smaller particles are pushed toward the top). This structure reduces the effect of channeling. Regeneration is the next step. The regeneration amount is usually given in terms of an acid or base concentration (6 mol% HCl, for example) at a prescribed flowrate for a minimum contact time. After regeneration, there are two rinse cycles. A slow rinse to remove excess regenerant from the bed followed by a fast rinse. The volume of these regenerant fluids is often a major cost for operation and disposal.

Capacity loss can arise from issues listed above, fouling, resin modification, etc. The flow distribution for the column will also affect the bed loading until breakthrough. Any variation in flow distribution can lead to channeling and premature breakthrough.

Some typical ion-exchange gel-type resins and their physical properties are described in Table 8.1.

The four different types of synthetic ion-exchange resins are: strong acid, weak acid, strong base, and weak base. Acidic resins have negative fixed charges and can exchange cations; basic resins have positive fixed charges and can exchange anions. Strong resins are fully ionized and all the fixed groups are available to exchange counterions, while weak exchangers are only partially ionized at most pHs (this often results in a lower exchange capacity, but makes regeneration easier) [4].

Strong-acid exchangers

Strong-acid exchangers easily remove all cations in solution. They have highly reactive sites such as the sulfonic group ($-SO_3H$), phosphonic group (H_2PO_3-), or hydroxyl group ($OH-$) [4]. Benzene–sulfonic acid groups (Figure 8.2) on a polystyrene–DVB polymer is the most common strong-acid resin; it is essentially an immobilized acid [6].

The commercial resins have dry weight capacities of 5.0 ± 0.1 eq/kg (typical wet-exchange capacity of $c_{RT} = 2.0$ eq/L, although this number varies as the degree of swelling changes). These resins are very stable and commonly have 20 or more years of service [6].

Strong-acid exchangers can be operated in a hydrogen cycle (often used for water demineralization) or in a sodium cycle (used for water softening). These cycles consist of

Table 8.1 *Properties of macroporous resins [6]. Reproduced with kind permission of Kluwer Academic Publishers.*

Resin	Bulk density, $\rho_{b,wet}$ (drained), kg/L	% Swelling due to exchange	Max T, °C	pH range	Wet-exchange capacity, eq/L	Max flowrate, m/h	Regenerant
Polystyrene–sulfonic acid							HCl, H₂SO₄, or NaCl
4% DVB	0.75–0.85	10–12	120–150	0–14	1.2–1.6	30	
8–10% DVB	0.77–0.87	6–8		0–14	1.5–1.9	30	
Polyacrylic acid (gel)	0.70–0.75	20–80	120	4–14	3.3–4.0	20	110% of theory HCl, H₂SO₄
Polystyrene–quaternary ammonium	0–7	~20	60–80	0–14	1.3–1.5	17	NaOH
Polystyrene–tert-amine (gel)	0.67	8–12	100	0–7	1.8	17	NaOH

Figure 8.2 Strong-acid ion-exchange monomer (benzene–sulfonic acid) [6]. Reproduced with kind permission of Kluwer Academic Publishers.

two steps (R^- represents the negatively charged functional group of the resin), the second of which is the regeneration step. The hydrogen cycle can be regenerated with HCl or with H₂SO₄, and the sodium cycle can be regenerated with NaCl [4].

Hydrogen cycle:
$$\begin{cases} CaSO_4 + 2(R^-H^+) \rightarrow (2R^-)Ca^{2+} + H_2SO_4 \\ (2R^-)Ca^{2+} + H_2SO_4 \rightarrow CaSO_4 + 2(R^-H^+) \end{cases}$$

Sodium cycle:
$$\begin{cases} CaSO_4 + 2(R^-Na^+) \rightarrow (2R^-)Ca^{2+} + Na_2SO_4 \\ (2R^-)Ca^{2+} + 2NaCl \rightarrow 2(R^-Na^+) + CaCl_2. \end{cases}$$

Strong-acid exchangers can also convert neutral salts into their corresponding acids if operated in the hydrogen cycle which is a process known as salt splitting (weak-acid resins cannot do this) [4].

In order for the resin to maintain charge neutrality, the fixed R^- group will attract counterions (H^+ and Na^+ in the previous example reactions) and it is these groups which will exchange with another ion in solution. The resin tends to prefer ions of higher valence, and also ions that are smaller and can more easily enter the resin pore structure. The preference series for the most common cations is [7]:

$$Ba^{2+} > Pb^{2+} > Sr^{2+} > Ca^{2+} > Ni^{2+} > Cd^{2+} > Cu^{2+} > Co^{2+} > Zn^{2+} > Mg^{2+}$$
$$> Ag^+ > Cs^+ > K^+ > NH_4^+ > Na^+ > H^+.$$

Since H^+ is the least preferred, the second step of the hydrogen cycle will require a large excess of acid before the concentration driving force is sufficient to favor regeneration; 60–75% of the regenerant can often go unused [4]. As stated earlier, this is an important cost and disposal consideration when using this technology.

Weak-acid exchangers

Weak-acid cation-exchange resins have weak fixed reactive sites such as the carboxylic group ($-COOH$). They are usually copolymers of DVB and acrylic or methacrylic acid [6]. The polyacrylic acid is shown in Figure 8.3, and properties are listed in the previous Table 8.1.

The weak-acid resins have a large number of acid groups that will only be partially ionized. Unlike strong-acid resins, these resins are not useful at low pHs (4 or 5) since the functional groups are not ionized and their effective capacity is zero under these conditions. Weak-acid resins also swell much more than strong-acid resins (sometimes as much as a 90% increase in volume when H^+ is replaced by the much larger Na^+ ion) which can cause excessive pressure drop, resin rupture, and equipment breakage. They are chemically stable, but may break from repeated swelling and shrinking cycles, and they are also "tight" compared to the strong-acid exchangers, so that the mass transfer resistances inside the resin are high and the resulting mass transfer zone in the bed is long [6].

Weak-acid resins can quickly remove cations from weak bases (like Ca^{2+} and Mg^{2+}), but they are not efficient at removing cations from strong bases (like Na^+ and K^+), and

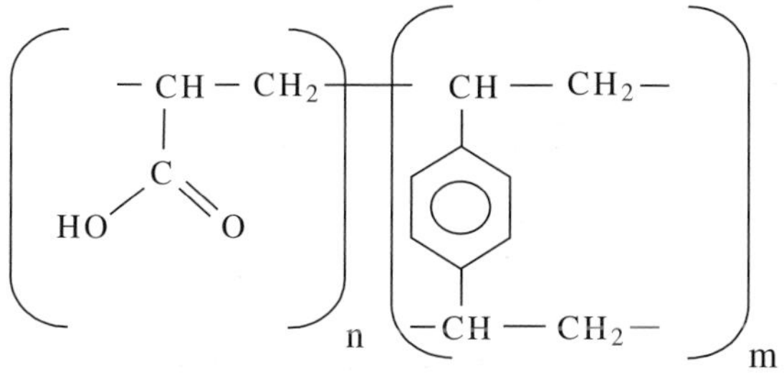

Figure 8.3 Weak-acid ion-exchange monomer (polyacrylic acid with DVB crosslink) [6]. Reproduced with kind permission of Kluwer Academic Publishers.

they can split only alkaline salts (e.g., $NaHCO_3$ but not $NaCl$ or Na_2SO_4). The preference series for common ions is similar to the strong-acid exchangers, but the H^+ position is moved to the left, sometimes as far as Ag^+ (and therefore, weak-acid resins are easier to regenerate than strong-acid resins) [7]. Carboxylic functional groups have such a high affinity for H^+ that they can use up to 90% of the acid (HCl or H_2SO_4) regenerant, even with low acid concentrations [4].

Weak-acid exchangers do not require as high a concentration driving force as strong-acid exchangers do. They do require an alkaline species to react (carbonate, bicarbonate, or hydroxyl ion):

$$Ca(HCO_3)_2 + 2(R^-H^+) \rightarrow (2R^-)Ca^{2+} + 2(H_2CO_3). \tag{8.5}$$

The regeneration step can be performed with HCl or H_2SO_4. These ion-exchange resins are often used for simultaneous softening and dealkalization in water treatment, and they are favored when the untreated water is high in Ca^{2+} and Mg^{2+} but low in dissolved CO_2 and Na. Sometimes a weak-acid exchange process is followed with a strong-acid exchange polishing step to minimize the higher cost of the strong-acid exchange process [4].

Strong-base exchangers

The two most common strong-base resins are also based on polystyrene–DVB polymers [6]. They have fixed reactive sites that are derived from quaternary ammonium groups (Figure 8.4). Type I has a greater chemical stability, but Type II has a higher regeneration efficiency and higher capacity. Both of these resins are fully ionized and are essentially equivalent to sodium hydroxide [4]. Typical wet-exchange capacities are in the range of

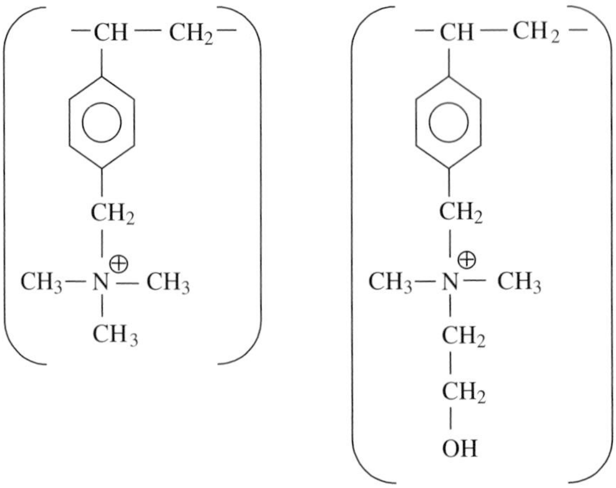

Figure 8.4 Strong-base ion-exchange monomer (quaternary ammonium structures). Type I is shown on the left and Type II is on the right [6]. Reproduced with kind permission of Kluwer Academic Publishers.

$c_{RT} = 1.0$ to 1.4 eq/L, and the resins can degrade, particularly at temperatures above 60 °C [6].

These resins work well over all pH ranges, and readily remove all anions. The preference series for the most common anions is [8]:

$$SO_4^{2-} > I^- > NO_3^- > CrO_4^{2-} > Br^- > Cl^- > OH^-.$$

Like the strong-acid exchangers, they can split neutral salts into their corresponding bases via the hydroxide cycle, and they are also often used in a chloride cycle to remove nitrates and sulfates from municipal water supplies. The hydroxide cycle is regenerated with a strong base like NaOH, while the chloride cycle is regenerated with NaCl [4].

Hydroxide cycle:
$$\begin{cases} NaCl + R^+OH^- \rightarrow NaOH + R^+Cl^- \\ NaOH + R^+Cl^- \rightarrow NaCl + R^+OH^- \end{cases}$$

Chloride cycle:
$$\begin{cases} NO_3^- + R^+Cl^- \rightarrow R^+NO_3^- + Cl^- \\ R^+NO_3^- + NaCl \rightarrow R^+Cl^- + NaNO_3. \end{cases}$$

Even weakly ionized substances, like silica and CO_2, can be removed with strong-base exchangers. Sometimes these exchangers are used after a cation exchanger for complete water demineralization [4].

Weak-base exchangers

A common type of weak-base exchanger uses the same polystyrene–DVB polymer but contains tertiary amine groups (Figure 8.5) [6].

The weak-base resins are fully ionized at low pHs and not ionized at all at high pH [5]. Like the weak-acid resins, they can suffer from oxidation and organic fouling.

Weak-base exchangers are able to remove anions from strong acids like Cl^-, SO_4^{2-}, and NO_3^-. They do not remove anions from weak acids very well (silicic, HCO_3^-,

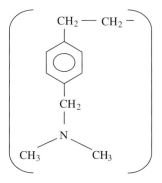

Figure 8.5 Weak-base ion-exchange monomer (tertiary amine on polystyrene) [6]. Reproduced with kind permission of Kluwer Academic Publishers.

Table 8.2 *Particle size in US mesh values and in mm [4].*

US mesh	Diameter (mm)
16–20	1.2–0.85
20–50	0.85–0.30
50–100	0.30–0.15
100–200	0.15–0.08
200–400	0.08–0.04

CO_3^{2-}, and SiO_4^{2-}). The preference series for some common anions is similar to that for strong-base exchangers, but OH^- will fall farther to the left depending on the strength of the reactive group on the resin [8]. They can be regenerated with $NaOH$, NH_4OH and Na_2CO_3, and they have much higher regeneration efficiencies than strong-base exchangers. Weak-base exchangers are sometimes used in conjunction with strong-base exchangers to minimize cost and to attract organics that might otherwise foul the strong-base resins [4].

8.5.1 Physical characteristics of the resins

Most synthetic resins are granular with a spherical diameter of 0.04–1.0 mm. In the United States, the particle sizes are listed according to standard screen sizes or "mesh" values. Table 8.2 shows a comparison of mesh sizes and metric sizes; the most common size ranges used in large-scale applications are 20–50 and 50–100 mesh [4].

The particle size has a significant effect on the hydraulics of an ion-exchange column. In about half of all ion-exchange applications, the design is based on hydraulic rather than chemical limitations (the allowable pressure drop dictates the smallest particle size used) [4]. The size of the resin particles also affects the kinetics of ion exchange; the rate of exchange can be proportional to the inverse of the diameter or the inverse of the square of the diameter (kinetics are discussed later). This tradeoff between hydraulic and transport considerations with respect to particle size is again similar to the discussion in Chapter 7: Adsorption.

8.6 Equilibria

Ion exchange is a chemical reaction, and in theory will continue until equilibrium is reached between the bulk solution and the solution within the pores of the resin. However, in most real processes equilibrium is not achieved and the effluent solution from an ion-exchange column is not in equilibrium with the influent solution, so kinetics as well as equilibrium are needed to describe the ion-exchange process. Equilibrium can be useful to predict the maximum amount of separation that can be achieved for a particular solution and resin.

8.6.1 Applying the law of mass action

The generalized form of Equation (8.2) for cation exchange between two ions (one (A) in solution, and one (B) originally in the resin) is:

$$z_A \overline{B}^{z_B^+} + z_B A^{z_A^+} \leftrightarrow z_B \overline{A}^{z_A^+} + z_A B^{z_B^+},$$ (8.6)

where z_A^+ and z_B^+ are the valence and charge of ions A and B. Since this reaction is reversible, we can write an expression for the thermodynamic equilibrium constant:

$$K_a = \frac{(\overline{a}_A)^{z_B}(a_B)^{z_A}}{(\overline{a}_B)^{z_A}(a_A)^{z_B}}.$$ (8.7)

It is often difficult to determine the activity coefficients for the resin phase while the activity coefficients are approximately 1 for dilute solutions. So, a selectivity coefficient K_c is often used:

$$(K_c)_B^A = \frac{(\overline{C}_A)^{z_B}(C_B)^{z_A}}{(\overline{C}_B)^{z_A}(C_A)^{z_B}}.$$ (8.8)

For a binary system, this coefficient can be written in terms of ion fractions

$$(K_c)_B^A \left(\frac{\overline{C}}{C}\right)^{z_A - z_B} = \frac{(y_A)^{z_B}(x_B)^{z_A}}{(y_B)^{z_A}(x_A)^{z_B}}.$$ (8.9)

A distribution factor can be defined as

$$m_A = \frac{\overline{C}_A}{C_A} = \frac{y_A \overline{C}}{x_A C}.$$ (8.10)

The separation factor α is:

$$\alpha_{AB} = \frac{y_A/x_A}{y_B/x_B} = \frac{y_A x_B}{y_B x_A},$$ (8.11)

where y represents the resin phase and x represents the solution (fluid) phase.

For univalent exchange, Equation (8.2), $z_A = z_B = 1$. Equation (8.8) becomes

$$(K_c)_B^A = \frac{y_A x_B}{y_B x_A} = \alpha_{AB}.$$ (8.12)

Univalent–divalent exchange ($z_A = 2$, $z_B = 1$) is an important application of ion-exchange technology. For this case, the selectivity coefficient is:

$$(K_c)_B^A \left(\frac{\overline{C}}{C} \right) = \frac{y_A(1 - x_A)^2}{x_A(1 - y_A)^2}. \qquad (8.13)$$

The preference for A (divalent ion) is greatly increased as the solution concentration is decreased. As one increases C, the selectivity will decrease.

8.6.2 Multicomponent ion exchange

The theoretical treatment of multicomponent ion exchange involving species of different valences is extremely complex. For ternary systems, the theory is reasonable and quite precise, although the representation relies heavily on the constancy of the relevant separation factors. Rigorous treatments are available [8–10]. Regrettably, the values of the selectivity coefficients and separation factors are not constant in practice. More detailed information on representing and predicting multicomponent ion exchange can be found in [11–14].

8.7 Equipment and design procedures

8.7.1 Equipment

Because the ion-exchange process is very similar to the adsorption process, the equipment is also similar. Ion exchange occurs in batch tanks, stirred tanks, fixed-bed columns, fluidized-bed columns, and moving-bed processes. The most common process is a fixed-bed column (Figure 8.6), which can be operated in a cocurrent or countercurrent fashion.

Both cocurrent and countercurrent fixed-bed columns operate in the same way during ion exchange, but the regeneration steps are different. For a cocurrent column, after the column reaches a predetermined breakthrough level, it is removed from the process and regenerated by passing the regenerate solution through the column in the same flow direction as the solution that is being treated. A countercurrent column will have the regenerate solution pass through in the opposite flow direction from the solution being treated. Cocurrent columns are simpler to design and operate, while countercurrent columns have higher chemical efficiencies [4].

Moving-bed columns are based on a moving packed bed in counterflow to the normal service flow of solution. It is always necessary to shut off normal flow to achieve bed movement, and hydraulic displacement is used to move the bed with either a piston or pump driving the displacement water [1]. The two most successful commercial designs are the Higgins technique and the Asahi moving packed bed (Figures 8.7 and 8.8).

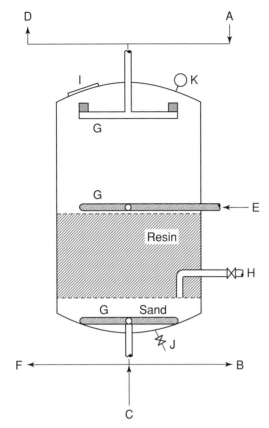

Figure 8.6 Typical fixed bed. A, feed; B, effluent; C, backwash supply; D, backwash overflow to resin trap; E, eluant supply; F, spent eluant; G, distributor manifolds; H, resin removal line; I, access hole; J, drain; K, pressure gage and vent [1]. Copyright 1987 John Wiley & Sons, Inc. This material is used by permission of John Wiley & Sons, Inc.

Mixed beds contain a thoroughly mixed bed of strong-acid and strong-base resins. These beds are used for deionization of water and typically provide better water quality than two beds in series. A key issue is regeneration. The two resins need to be separated after the ion-exchange cycle and prior to regeneration. Strong-base resins are typically lighter than strong-acid resins. Backwashing can be used to segregate the resins. The regeneration is then done simultaneously with the appropriate regenerant introduced at the top and bottom of the column and withdrawn at the region between the two segregated resins. Obviously, the resins will need to be remixed prior to the next ion-exchange cycle.

8.7.2 Design procedures

Many of the design principles that apply to adsorption columns may also be applied to ion-exchange columns. For example, breakthrough curves from a pilot column along

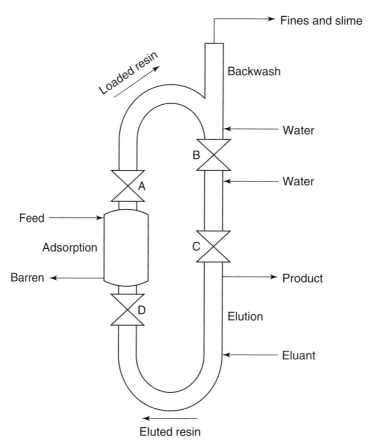

Figure 8.7 Higgins moving packed bed [1]. Copyright 1987 John Wiley & Sons, Inc. This material is used by permission of John Wiley & Sons, Inc.

with an equation that describes kinetics may be used to design a full-size ion-exchange column. The kinetic equation given for an adsorption column design will often apply to ion exchange also [7]:

$$\ln\left(\frac{C_0}{C} - 1\right) = \frac{k_1 q_0 M}{Q} - \frac{k_1 C_0 V}{Q},\tag{8.14}$$

where C = effluent solute concentration

C_0 = influent solute concentration

k_1 = rate constant

q_0 = maximum solid-phase concentration of the solute (g/g or lb/lb)

M = mass of the resin

V = throughput volume

Q = flowrate.

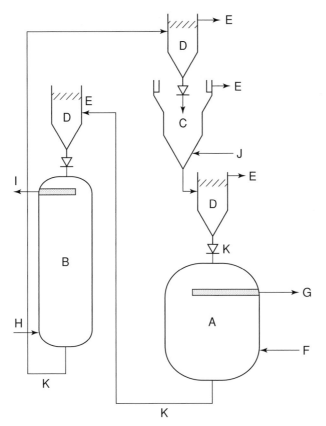

Figure 8.8 Asahi moving packed bed. A, adsorption section; B, elution section; C, fluidized resin backwash; D, resin collection hoppers with screen top and non-return valve outlet for resin; E, transfer and backwash water overflow; F, feed; G, barren effluent; H, eluant; I, eluate product; J, backwash supply; K, resin flow [1]. Copyright 1987 John Wiley & Sons, Inc. This material is used by permission of John Wiley & Sons, Inc.

When a pilot column is to be used to determine the breakthrough data, it is important to remember that it should be operated at the same flowrate as the full-size column in terms of bed volumes per hour.

Example 8.1: ion exchange in waste treatment

Problem:

An industrial wastewater with 100 mg/L of Cu^{2+} (3.2 meq/L) is to be treated by an ion-exchange column. The allowable effluent concentration, C_a, is 5% of C_0. A breakthrough curve has been obtained from an experimental laboratory column on the sodium cycle. Data concerning the column are: mass of resin = 41.50 g on a moist basis (23.24 g on a dry basis), moisture $= 45\%$, bulk density of resin $= 750 \text{ kg/m}^3$ on moist basis, and liquid flowrate $= 1.0 \text{ L/day}$. The design column will have a flowrate of 400,000 L/day, the allowable breakthrough time is seven days of flow, and the resin depth is approximately ten times the column diameter. Using the kinetic approach to column design, determine:

(a) The mass of resin required;
(b) the column diameter and depth.

Data from breakthrough test.

Volume (L)	C (mg/L)	C (meq/L)
15.9	4.5	0.14
18.1	17.2	0.54
19.5	40.0	1.26
20.7	62.9	1.98
22.0	86.4	2.72
23.4	98.2	3.09

Solution:

A plot of $\ln(C_0/C - 1)$ vs V gives a straight line with a slope of $-k_1 C_0/Q$ and an intercept of $k_1 q_0 M/Q$. The plot of the data is shown in Figure 8.9.

$$k_1 = (\text{slope})(Q/C_0)$$

$$= (0.76/\text{L})(1.0 \text{ L/day})(\text{L}/3.2 \text{ meq})(1000 \text{ meq/eq}) = 240 \text{ L/day} \cdot \text{eq}.$$

$$q_0 = (\text{intercept})Q/(k_1 M)$$

$$= (15.3)(1.0 \text{ L/day})(\text{day} \cdot \text{eq}/240 \text{ L})(1/23.24 \text{ g}) = 2.74 \times 10^{-3} \text{ eq/g}.$$

The mass of the resin in the design column can now be calculated:

$$\ln\left(\frac{C_0}{0.05\,C_0}\right) = \frac{(240 \text{ L/day} \cdot \text{eq})(2.74 \times 10^{-3} \text{ eq/g})M}{(4.0 \times 10^5 \text{ L/day})}$$

$$- (240 \text{ L/day} \cdot \text{eq})(3.2 \text{ meq/L})(1 \text{ eq}/10^3 \text{ meq})(7 \text{ days}).$$

So, (a) $M = 5.1 \times 10^6 \text{ g} = 5.1 \times 10^3$ kg resin (dry weight).

This amount of resin has a volume of:

$$\underbrace{\text{dry} \rightarrow \text{moist}}\qquad \overset{\text{moist density}}{\swarrow}$$

$$(5.1 \times 10^6 \text{ kg})(1/0.55)(\text{m}^3/750 \text{ kg}) = 12.4 \text{ m}^3.$$

The diameter, D, is calculated from $(\pi/4)(D^2)(10\,D) = 12.4 \text{ m}^3$.

So, (b) $D = 1.2$ m, and the depth $Z = 10\,D = 12$ m.

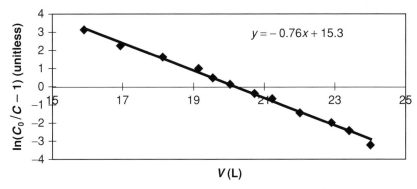

Figure 8.9 Ion exchange in waste treatment, Example 8.1.

Example 8.2 [15]

Problem:

A certain ion-exchange resin used for treating wastewater contains a finite quantity of charged groups. Therefore, the equilibrium can be expressed in the same way that an adsorption equilibrium is described: with an isotherm. Laboratory analysis of this resin shows that it follows the Langmuir isotherm:

$$Y = \frac{aX}{b + X},$$ (8.15)

where Y = amount exchanged (mass contaminant/volume resin)

X = concentration in solution (mass contaminant/volume water)

$a = 70$ mg/cm^3

$b = 50$ mg/L.

(a) If you have 1.5 L of an aqueous waste stream containing 220 mg/L contaminant, how much fresh resin is necessary to adsorb 90% of the contaminant? Solve this part algebraically.

(b) How many countercurrent stages would be required to adsorb 95% of the contaminant in the same feed as part (a) using 6.25 cm^3 of pure resin? Solve this part graphically.

Solution:

(a) A schematic diagram is shown in Figure 8.10.

Contaminant balance: $1.5(220) = SY_1 + 1.5(22)$.

$$\therefore S = 297/Y_1.$$

Equilibrium: $Y_1 = 70X_1/(50 + X_1)$ where Y_1 is in mg/cm^3 and X_1 is in mg/L.

At $X_1 = 22$ mg/L $\Rightarrow Y_1 = 21.4$ mg/cm^3 resin.

Plugging into contaminant balance, $S = 13.9\ \text{cm}^3$ resin.

Figure 8.11 is a schematic of the number of stages.

(b) To arrive at a graphical solution, Figure 8.12, use:

Overall mass balance: $1.5(220) = 1.5(11) + 6.25Y_1$

$$Y_1 = 50.2\ \text{mg/cm}^3.$$

Then, 1 plot equilibrium data using Langmuir isotherm;
2 plot operating line: (X_n, Y_{n+1}) and (X_0, Y_1);
3 determine the step off stages.

The number of stages required is three.

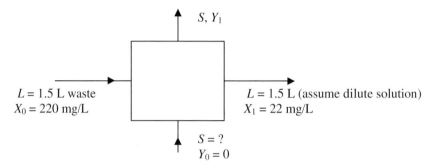

$L = 1.5\ \text{L waste}$
$X_0 = 220\ \text{mg/L}$

$S,\ Y_1$

$L = 1.5\ \text{L (assume dilute solution)}$
$X_1 = 22\ \text{mg/L}$

$S = ?$
$Y_0 = 0$

Figure 8.10 Schematic for wastewater treatment, Example 8.2.

$X_0 = 220,\ L = 1.5$

1

...

n

$X_n = 11,\ L = 1.5$

$S = 6.25,\ Y_1 = ?$

$S = 6.25\ \text{cm}^3,\ Y_{n+1} = 0$

Figure 8.11 Number of countercurrent stages, Example 8.2.

8.8 Remember

- Ion exchange is a UNIT OPERATION. Regardless of what chemicals are being separated, the basic design principles for ion exchange are always similar.
- An ion-exchange process will look much like an adsorption process, with a liquid phase being contacted with a solid phase (the RESIN) where the ion exchange occurs.
- Like adsorption, an ion-exchange process can be carried out in a fixed-bed column, or in a batch process. Columns are more appropriate for industry and water or wastewater applications, while batch processes are more likely to be used in a lab for analytical purposes.
- An ion-exchange process can remove anions or cations from solution, depending on the charge of the functional group within the resin.
- There are some naturally occurring ion-exchange resins as well as synthetic resins.

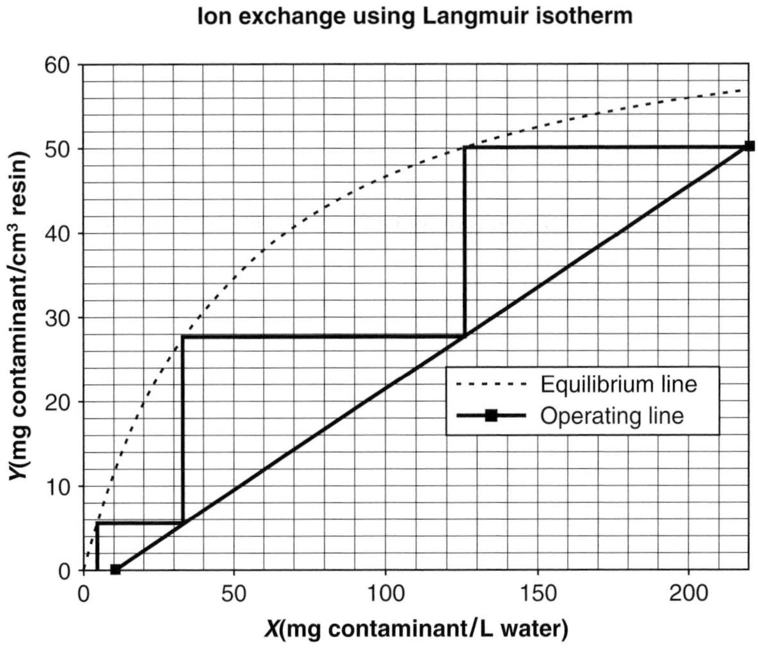

Ion exchange using Langmuir isotherm

Figure 8.12 Graphical solution, Example 8.2(b).

- Ion exchange is a reversible process, and the resin can be regenerated. However, a resin can be fouled and is susceptible to physical damage, so lifetime is an important issue.

8.9 Questions

8.1 Describe the difference between a weak and a strong ion-exchange material.

8.2 How could the pH of the fluid to be treated affect the capacity of an ion-exchange resin?

8.3 What is the primary difference between ion exchange and absorption?

8.4 What is required to completely de-mineralize a water stream through ion exchange?

8.10 Problems

8.1 For the test column and breakthrough curve given in Example 8.1, determine the meq of Cu^{2+} ion removed per 100 grams of resin on a dry weight basis at the allowable breakthrough volume, V_{bt}, for $C_a = 0.05\ C_0$. Also, determine the meq of Cu^{2+} ion removed per 100 grams of resin on a dry weight basis at complete exhaustion. The dry weight of resin used was 23.24 grams.

9

Membranes

A wide screen just makes a bad film twice as bad.

– SAMUEL GOLDWYN

9.1 Objectives

1 Identify the three major driving forces for membrane separations.
2 Define permeability, permeance, selectivity, and rejection.
3 List the transport mechanisms for membrane separations.
4 List some environmental applications of each type of membrane separation.
5 Describe the advantages and disadvantages of membrane technology.

9.2 Membrane definition

A membrane can be defined as [1]:

...a semi-permeable barrier between two phases. This barrier can restrict the movement of molecules across it in a very specific manner. The membrane must act as a barrier between phases to prevent intimate contact. This barrier can be solid, liquid, or even a gas. The semi-permeable nature is essential to insuring that a separation takes place. If all species present could move through the membrane at the same rate, no separation would occur. The manner in which the membrane restricts molecular motion can take many forms. Size exclusion, differences in diffusion coefficients, electrical charge, and differences in solubility are some examples. A membrane separation is a rate process. The separation is accomplished by a driving force, not by equilibrium between phases.

There are three important points to note with respect to this definition. First, a membrane is defined by what it does (function), not by what it is. So, a wide range of materials are potentially useful as membranes. Second, the membrane separation mechanism is not specified. So, again there could be several choices. Third, there are four general types of membranes, classified by their material structure: polymers, liquids, inorganic and ion-exchange resins. Polymer membranes are the most common synthetic membrane. Some types of polymer membranes are: asymmetric skin (reverse osmosis), homogeneous films (used in the food and packaging industries), and composites (including polymer blends and actual layered membranes). Liquid membranes include supported films (which can accomplish both liquid- and gas-phase separations), emulsions (liquid-phase separations), and bulk membranes (not used commercially). Inorganic membranes can be ceramic, glass or metal; the newer ceramic membranes are proving to be useful in high-temperature and corrosive environments. Ion-exchange membranes have a polymeric backbone with fixed charge sites (cationic or anionic). These membranes are used in chlor–alkali industry and fuel cells. They are also used as supports for carrier-impregnated liquid membranes.

Synthetic membranes can further be classified as symmetric or asymmetric (Figure 9.1) [2]. The thickness of symmetric membranes (porous or non-porous) ranges roughly from 10 to 200 μm. The resistance to mass transfer in the membrane is inversely proportional to the membrane thickness. A decrease in membrane thickness results in an increased permeation rate.

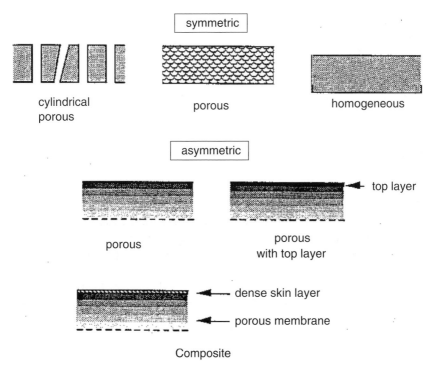

Figure 9.1 Schematic representation of membrane cross-sections [2]. Reproduced with kind permission of Kluwer Academic Publishers.

A breakthrough to industrial applications was the development of asymmetric polymer membranes. These consist of a very dense top layer or skin with a thickness of 0.1 to 0.5 μm supported by a porous sublayer with a thickness of about 50 to 150 μm. These membranes combine the high selectivity of a dense membrane with the high permeation rate of a very thin membrane.

It is also possible to obtain composite membranes. The top layer and support layer originate from different materials; each layer can be optimized independently. The top layer provides the separation while the support layer provides mechanical support. The support layer should <u>not</u> be a significant mass transfer resistance.

9.3   Pluses and minuses for membrane processes

<u>Pluses</u>
- Uses energy as separating agent.
- Can separate materials from molecular size up to particle size.
- Generally has low energy use since no phase change occurs (except for pervaporation). This is particularly true for liquid systems. Gas systems may or may not use large amounts of energy.
- Can be good for economically removing small amounts of materials even when they are not selectively permeated.
- Generally has a very simple flowsheet.
- Compact design, so little space is required.
- Can have a high separation factor in many cases.
- In some cases can interface well with other separation processes to form hybrids.
- Scales down well for small applications.
- Avoids damage to products since it is normally an ambient temperature process.

<u>Minuses</u>
- Often can only be used for concentrating a product as a retentate instead of producing two relatively pure products. This is especially true for reverse osmosis separations.
- Can have chemical incompatibilities between membrane materials and process streams. This is a difficult problem in the chemical and petroleum industries.
- Often cannot operate at much above room temperature. Ceramic and metal membranes can expand the temperature range of operation.
- Often does not scale-up well to accept massive flows. There is no economy of scale. Membrane units are modular. For twice the throughput, you need twice the number of modules.
- Membrane fouling, especially with liquid-phase feeds, causes flux decline and can reduce lifetime.
- Can require chemicals to be added to the feed to adjust pH or reduce fouling.

• Equipment required to produce required high pressures for some applications can be extremely noisy.

9.4 Environmental applications

Though still a relatively new commercial process in comparison with other technologies in this text, membrane separation processes should increase in applications in the future. In the near term, air and water treatment applications probably represent the best opportunities for membranes – though there are important additional environmental applications. Below is a list of various membrane processes and their environmental applications.

Gas separations

Carbon-dioxide removal from various gas streams.

Acid-gas (CO_2, H_2S) separation from natural gas.

Drying of gas streams. One important example is dehydration of natural gas. Eliminates use of solvents for this application.

Removal of organic compounds from vent streams. Recovery of ethylene and propylene from vent streams from polyethylene and polypropylene pellet storage is one commercial example.

Reverse osmosis (RO) and nanofiltration (NF)

Reduction of chemical oxygen demand (COD) of wastewater and groundwater streams.

Color removal from wastewater streams.

Removal of various ions from wastewater streams.

Clean-up of wastewaters from electroplating baths, landfill leachates and laundry effluents.

Concentration of spent sulfite liquor from paper-plant effluents.

Recovery of homogeneous catalysts.

Ultrafiltration (UF)

Concentration of latex particles in water and recovery of latex particles from wastewaters (painting processes, for example).

Removal of polymer constituents from wastewaters.

Separation of oil–water emulsions.

Microfiltration (MF)

Separation of oil–water emulsions.

Removal of precipitated metal hydroxides.

Removal of micron-size particles from a wide variety of liquid streams.

Concentration of fine solids.

Pervaporation

Removal of small amounts of water from organic solutions, e.g., water from isopropanol.

Removal of small amounts of organics from water, e.g., in wastewater clean-up.

Separate liquid-phase solutions that form an azeotrope. Replaces azeotropic distillation which eliminates use of solvents.

Electrodialysis (ED)

Removal of heavy metals, nitrates, and cyanides from water streams.

Desalination of brackish water.

Processing of rinse waters in the electroplating industry.

9.5 Basic parameters and separation mechanisms

The mass transport across the membrane is caused by one or more driving forces. A gradient in chemical or electrical potential is the usual basis for these driving forces. Pressure or concentration differences are the most common driving forces. The total flux (N_i) of a species is related to the driving force (ΔP or ΔC) by

$$N_i = \frac{Q}{l}(\Delta P). \tag{9.1}$$

The driving force is ($\Delta P - \Delta \Pi$) when there is an osmotic pressure difference ($\Delta \Pi$) across the membrane. Q is the permeability and Q/l is the permeance. Permeance is used when the actual membrane thickness is unknown. The separation factor α_{ij} is calculated as the ratio of permeabilities or permeances.

Liquid separations, which typically involve filtration, use a term called rejection (R). This term, with values between 0 and 1, measures the degree of rejection (lack of permeation) for a given solute. It is defined as:

$$R = 1 - \frac{C_p}{C_f}, \tag{9.2}$$

where C_p = permeate concentration

C_f = feed concentration.

A value of 1 implies complete rejection.

For membrane separation processes, the mass-separating agent is the membrane itself. Energy is supplied to provide the driving force for permeation through the membrane. The mechanisms by which membranes separate different gas-phase components are outlined in Section 9.7. For liquid-phase separations, filtration is based on size. Larger particles are prevented from permeating while smaller components (typically solvents such as water) permeate. This mechanism is similar to molecular sieving and is equivalent if the degree of rejection is 1 (complete rejection). For pervaporation, a phase change is involved in liquid-phase separations.

9.6 Dense membranes

Dense membrane materials are usually polymers. They are usually classified as rubbery (amorphous) (above their glass transition temperature, T_g) or glassy (crystalline) (below T_g). As a polymer is cooled below T_g, it does not immediately become a completely crystalline material. Instead, there are both crystalline and amorphous regions. These two regions give rise to "dual mode" transport. The gas concentrated in the rubbery region (C_D) is characterized by a Henry's Law type sorption while a Langmuir-type isotherm describes gas concentration in the crystalline region (C_H). The total gas concentration is written as the sum of the two contributions:

$$C = C_D + C_H \tag{9.3}$$

$$= k_D P + \frac{C_H' b P}{1 + b P}, \tag{9.4}$$

where k_D = Henry's Law type coefficient

$\quad C_H'$ = sorption capacity in crystalline region

$\quad b$ = constant

$\quad P$ = gas pressure.

The gas transport across the membrane is also the sum of the contributions through the two regions

$$N = -D_D \frac{dC_D}{dx} - D_H \frac{dC_H}{dx}. \tag{9.5}$$

The pure component permeability for a large feed pressure and a very small permeate pressure is

$$Q = k_D D_D \left[1 + \frac{F K}{1 + b P} \right], \tag{9.6}$$

where

$$F = \frac{D_H}{D_D} \tag{9.7}$$

$$K = -\frac{C_H' b}{k_D}. \tag{9.8}$$

For mixtures, the solute concentration and permeability are modified to account for the fact that some fraction of the sorption capacity in the crystalline region is occupied by each solute:

$$c_i = k_{Di} P + \frac{C_H' b_i P}{1 + \sum\limits_{j=1}^{n} b_j P} \tag{9.9}$$

$$Q = k_{Di} D_{Di} \left[1 + \frac{F_i K_i}{1 + \sum\limits_{i=1}^{n} b_i P} \right]. \tag{9.10}$$

Example 9.1

Problem:

Based on the dual-mode transport model, calculate the pure component and mixtures permeability of CO_2 and CH_4 using a polysulfone membrane. The gas pressure is 10 atm for each component.

	$k_D \left(\frac{\text{cc(STP)}}{\text{cm}^3 \cdot \text{atm}} \right)$	$D_D \left(10^8 \frac{\text{cm}^2}{\text{s}} \right)$	$C_H' \left(\frac{\text{cc(STP)}}{\text{cm}^3} \right)$	$b \ (\text{atm}^{-1})$	F
CO_2	0.664	4.40	17.90	0.326	0.105
CH_4	0.161	0.44	9.86	0.070	0.349

Solution:

Dual-mode model:

Pure-component calculation:

$$Q = k_D D_D \left[1 + \frac{KF}{1 + bP} \right] \qquad K = \frac{C_H' b}{k_D}$$

$$Q_{CO_2} = 0.664(4.40 \times 10^{-8}) \left[1 + \frac{\left(\frac{17.90(0.326)}{0.664} \right)(0.105)}{1 + 0.326(10 \text{ atm})} \right]$$

$$Q_{CO_2} = 3.55 \times 10^{-8} \frac{\text{cc(STP)}}{\text{cm} \cdot \text{s} \cdot \text{atm}}$$

$$Q_{CH_4} = 0.161(0.444 \times 10^{-8}) \left[1 + \frac{\left[\frac{(9.86)(0.070)(0.349)}{0.161} \right]}{1 + 0.070(10 \text{ atm})} \right]$$

$$Q_{CH_4} = 1.34 \times 10^{-9} \frac{\text{cc(STP)}}{\text{cm} \cdot \text{s} \cdot \text{atm}}$$

$$\alpha_{\text{ideal}} = \frac{Q_{CO_2}}{Q_{CH_4}} = 26.$$

Mixture calculation:

$$Q_i = k_{Di} D_{Di} \left[1 + \frac{K_i F_i}{1 + b_i P_i + b_j P_i} \right]$$

Note the two bP terms in the denominator – this takes into account the competitive nature of the transport.

$$Q_{CO_2} = 0.664(4.4 \times 10^{-8}) \left[1 + \frac{\frac{17.90(0.326)}{0.664}(0.105)}{1 + 0.326(10) + 0.07(10)} \right]$$

$$Q_{CO_2} = 3.465 \times 10^{-8} \frac{cc(STP)}{cm \cdot s \cdot atm}$$

Note this is slightly reduced from the pure component analysis.

$$Q_{CH_4} = 0.161(0.444 \times 10^{-8}) \left[1 + \frac{\frac{9.86(0.070)}{0.161}(0.349)}{1 + 0.070(10) + 0.326(10)} \right]$$

$$Q_{CH_4} = 9.30 \times 10^{-10} \frac{cc(STP)}{cm \cdot s \cdot atm}$$

This answer is also slightly lower than the pure component analysis.

$$\alpha_{mixture} = \frac{Q_{CO_2}}{Q_{CH_4}} = 37.$$

9.7   Porous membranes

Four types of diffusion mechanisms can be utilized to effect separation in porous membranes. In some cases, molecules can move through the membrane by more than one mechanism. These mechanisms are described below. Knudsen diffusion gives relatively low separation selectivities compared to surface diffusion and capillary condensation. Shape selective separation can yield high selectivities. The separation factor for these mechanisms depends strongly on pore-size distribution, temperature, pressure, and interactions between the solute being separated and the membrane surfaces.

9.7.1   Knudsen diffusion

Under viscous flow (Poiseuille flow), the mean free path of fluid molecules is small compared to the pore diameter, and molecules undergo many more collisions with each other than with the walls of the membrane. The molecules in a mixture do not behave independently in viscous flow and no separation is possible. Thus, viscous flow is not desirable. As the pressure is lowered, the mean free path (λ) of the molecules becomes longer than the pore diameter (Figure 9.2(a)). As a result, the molecules undergo far more collisions with the pore walls than with each other, and the molecules flow through the pores independently from each other. The mean free path of a gas molecule can be calculated as:

$$\lambda = \frac{kT}{\sqrt{2}\sigma P},$$
(9.11)

where k = Boltzmann's constant
$\quad T$ = absolute temperature
$\quad P$ = absolute pressure
$\quad \sigma$ = molar average collision diameter.

Table 9.1 *Mean free paths for various gases.*

Gas	σ (nm)	Mean free path (nm)					
		500 K			800 K		
		0.1 MPa	1.0 MPa	5.0 MPa	0.1 mPa	1.0 MPa	5.0 MPa
H_2	0.2915	183	18.3	3.7	293	29	5.9
CO	0.3706	113	11.3	2.3	181	18.1	3.6
N_2	0.3749	111	11.1	2.2	177	17.7	3.5
CO_2	0.3897	102	10.2	2.0	164	16.4	3.3

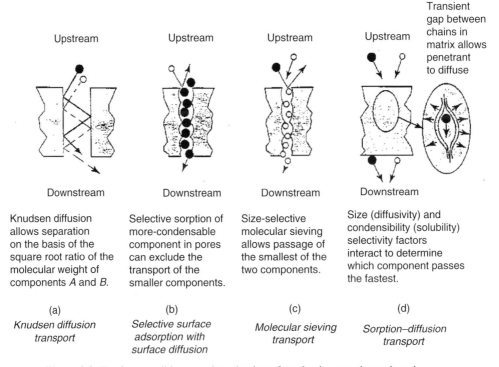

Upstream

Downstream

Knudsen diffusion allows separation on the basis of the square root ratio of the molecular weight of components *A* and *B*.

(a)
Knudsen diffusion transport

Upstream

Downstream

Selective sorption of more-condensable component in pores can exclude the transport of the smaller components.

(b)
Selective surface adsorption with surface diffusion

Upstream

Downstream

Size-selective molecular sieving allows passage of the smallest of the two components.

(c)
Molecular sieving transport

Upstream

Transient gap between chains in matrix allows penetrant to diffuse

Downstream

Size (diffusivity) and condensibility (solubility) selectivity factors interact to determine which component passes the fastest.

(d)
Sorption–diffusion transport

Figure 9.2 The four possible general mechanisms for selective-membrane-based gas and vapor separations [3]. Reproduced with permission of AIChE.

Table 9.1 indicates some representative values of the mean free path of several gases at various pressures and temperatures.

For a circular capillary of radius r and length l, the molar flux of component i in the Knudsen diffusion regime is given by:

$$J_i = \frac{8\pi r^3}{3\sqrt{2\pi M_i RT}} \frac{\Delta P_i}{l}, \tag{9.12}$$

where ΔP_i is the pressure drop of component i across the membrane and M_i is the molecular weight of component i. The flux through a microporous membrane of thickness l is given by:

$$J_i = \frac{GS}{\sqrt{2\pi M_i RT}} \frac{\Delta P_i}{l},$$ (9.13)

where G is a geometric factor that takes into account tortuosity and porosity. Note that the flux is independent of the average pressure as long as the pressure is in the Knudsen diffusion regime.

An equimolar mixture of feed gas, diffusing across a membrane in the Knudsen diffusion regime, will have a separation factor $\alpha_{ij} = \sqrt{M_j/M_i}$ when the permeate side is at vacuum. Otherwise, the separation factor will be smaller. The narrow pore-size distributions and the small pores of ceramic and glass membranes allow separation due to Knudsen diffusion (for the appropriate pressure range) by preferential diffusion of the lighter component through the membrane. In composite membranes, the thin permselective layer can be in the Knudsen diffusion regime and thus be responsible for all the separation. The support layers, with their larger-diameter pores, are usually in the viscous-flow regime.

Separation by Knudsen diffusion has some limitations because only the lighter component can be preferentially removed. The best separation in the Knudsen diffusion regime is thus obtained for H_2. When the molecular weight difference between components to be separated is small, an economical separation probably cannot be obtained by Knudsen diffusion.

9.7.2 Surface diffusion

A process that can occur in parallel with Knudsen diffusion is surface diffusion (Figure 9.2(b)). A gas can chemisorb or physisorb on the pore walls and migrate along the surface. Surface diffusion increases the permeability of the more strongly adsorbed components in a diffusing mixture while simultaneously reducing the permeability of the gas diffusing components by decreasing the effective pore diameter. Thus, this diffusion mechanism is more important for membranes with small pores. For example, the number of molecules in a monolayer on the wall of a 5-nm-diameter pore at 0.1 MPa can be over 200 times larger than the number of molecules in the gas phase. As the temperature increases, species desorb from the surface, surface diffusion becomes less important, and Knudsen diffusion predominates. When surface diffusion occurs, competitive adsorption must also be considered.

The equation describing surface flow is

$$J_s = -\rho(1 - \varepsilon)D_s\mu_s dq/dl$$ (9.14)

where J_s is the surface diffusion flux, ρ is the true density of the adsorbed layer, D_s is the surface diffusion coefficient, μ_s is the tortuosity of the surface, and dq/dl is the surface concentration gradient. Surface diffusion must usually be determined experimentally.

9.7.3 Capillary condensation

When one of the components in a mixture is a condensable vapor and the pores are small enough, the condensate can block gas-phase diffusion through the pores. This is the limiting case of surface diffusion where the adsorbed layer fills the pore. This condensate will evaporate on the low partial pressure side of the membrane. The Kelvin equation predicts that condensation can occur in small pores even through the partial pressure of that component is below its vapor pressure. The Kelvin equation represents thermodynamic equilibrium between the gas phase and fluid in the pore:

$$P/P_s = \exp\left[\frac{-2\gamma \cos \Theta}{\rho r\ RT}\right], \tag{9.15}$$

where P = vapor pressure in bulk phase in the presence of capillary pores

P_s = normal vapor pressure in the bulk phase

γ = surface tension of the condensed fluid in the pore

Θ = contact angle between the condensed fluid and the pore wall

ρ = molar density of the condensed liquid

r = mean pore radius.

As a result of capillary condensation, the pores can completely fill with that component (Figure 9.2(c)). For a narrow distribution of pore sizes, all pores will be filled and the fluxes of the other components through the membranes will be quite small and limited by their solubility in the condensable component. Thus, extremely high separation factors are possible.

9.7.4 Molecular-sieve separation

Molecular-sieve membranes can yield high separation factors by permitting small molecules to diffuse while essentially excluding or severely restricting the accessibility of larger molecules (Figure 9.2(d)). This type of diffusion, where the pores are of molecular size, has been referred to as shape selective or configurational diffusion.

9.8   Membrane configurations

Membrane processes operate in two basic modes. In Figure 9.3(a), the permeate stream is solely the components of the feed stream that transport across the membrane. Figure 9.3(b) illustrates the case where a sweep stream is introduced on the permeate side to collect the permeate. The sweep stream can operate cocurrent or countercurrent to the feed stream. One limiting case is when the streams on both sides of the membranes are perfectly mixed and there is no axial variation in solute concentration. These basic modes are then incorporated into various geometric configurations (Figure 9.4).

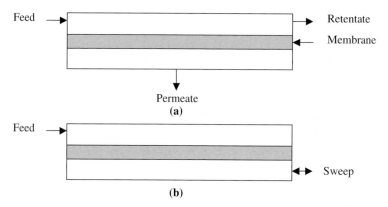

Figure 9.3 Basic modes for membrane processes.

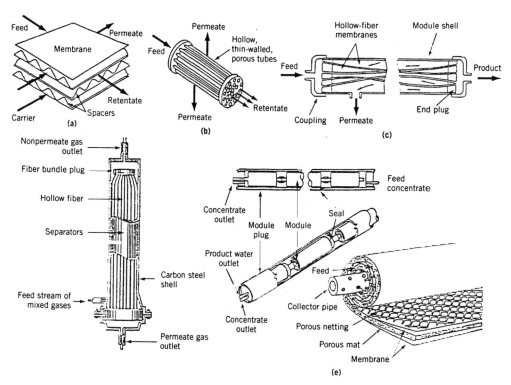

Figure 9.4 Schematic conventional membrane modules: (a) plate-and-frame, (b) tubular, (c) capillary, (d) hollow fiber (where the separator is 10–20 cm dia × 3–6 m long), and (e) spiral wound [6]. This material is used with the permission of John Wiley & Sons, Inc.

Membrane separating modules can be (a) flat sheets (such as continuous column, supported liquid, or polymer film), (b) tabular, (c) capillary, (d) hollow fibers (either coated fibers or supported liquid) and (e) spiral wound. The hollow fiber is similar to the tubular mounting except that hollow fibers typically have a much smaller diameter.

Table 9.2 *Existing and emerging applications for membrane vapor recovery.*

Existing	Emerging
Gasoline vapor recovery	Recovery of feedstock from oxidation reactor vents
Recovery of vinyl chloride monomer from PVC reactor vents	Recovery of olefins from reactor purge gas streams
Recovery of CFCs from process vent and transfer operations	Recovery of gasoline vapor from gas station storage tanks
Recovery of olefins from resin purge bin off-gas in polyolefin production	

In the spiral-wound mounting, a porous hollow tube is spirally wrapped with a porous sheet for the feed flow, and a membrane sheet and a porous sheet for the product flow to give a spiral sandwich-type wrapping. The spiral module is encased in a pressure vessel, and the feed flow through the porous sheet is in an axial direction to the porous tube. As the feed flow passes through the porous sheet, a portion of the flow passes through the membrane into the porous sheet for the product. From there, the product flows spirally to the porous center tube. The retentate stream is discharged from the downstream end of the porous sheet for the feed flow.

Single membrane units can be evaluated based on their geometry and operation conditions. Zolandz and Fleming [4] provide a good description for gas permeation systems and models for design purposes. Seader [5] discusses the use of cascades (or staging) for various series and/or parallel sets of membrane modules.

9.9   Membrane processes

The membrane processes described in this chapter are summarized in Table 9.2 [7], and will be discussed in more detail later in this chapter.

9.9.1   Gas separations and vapor permeation

For the separation of gas mixtures (permanent gases and/or condensable vapors) where the feed and permeate streams are both gas phase, the driving force across the membrane is the partial pressure difference. The membrane is typically a dense film and the transport mechanism is sorption–diffusion. The "dual-mode" transport model is typically used with polymer materials that are below their glass transition temperature.

The ratio of the permeate to feed flowrate is called the cut. For a given feed flowrate, the cut increases as the membrane area increases. The selectivity decreases as the cut increases. So, there is a tradeoff between productivity and selectivity in the design of a membrane unit for this application. This is illustrated in the Robeson log–log plot for

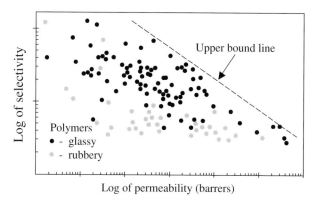

Figure 9.5 Robeson plot for polymer membranes.

polymer membranes (Figure 9.5). All polymers fall below an "upper bound" that decreases in selectivity as permeability increases. For permanent gas separations (CO_2/CH_4 and H_2S/CH_4), for example, the feed pressure is typically 1000 psi and the permeate pressure is near atmospheric. The high pressure provides a high flux but low selectivity. Thus, this process is generally used for bulk gas separations. The term vapor permeation is used when condensable vapors are present in the feed streams. An important example would be VOCs in air. There are two typical modes of operation. The feed is introduced to the membrane at atmospheric pressure and the permeate at vacuum. The condensable vapor is enriched in the permeate stream. Condensation and compression are then employed to transfer the permeate stream. The second mode employs compression and condensation of the feed stream prior to the membrane inlet. The permeate stream is operated at atmospheric pressure. Table 9.3 lists some existing and emerging applications for vapor permeation.

9.9.2 Pressure-driven filtration processes

The Filtration Spectrum is shown in Figure 9.6 [8]. A wide variety of materials and particle sizes are placed on the figure to give readers some perspective on the size range and materials separated by the various filtration processes. A detailed description of each process follows.

9.9.3 Microfiltration and ultrafiltration

As shown on the Filtration Spectrum, microfiltration membranes have pores in the range of 0.05 to 2 μm while ultrafiltration membranes are in the range of 0.003 to 0.1 μm. As the names imply, these processes are used to separate (filter) a solvent (usually water) from suspended solids or colloidal material. The solvent passes through the membrane while the solids are retained on the feed side of the membrane. The choice of membrane is dictated by the particle size of the retained material.

Table 9.3 *Major membrane processes [7].*

Process	Concept	Driving force	Species passed	Species retained
Microfiltration (MF)	Feed → / Retentate ↑ / Solvent ↑ / Microporous membrane	Pressure difference 100–500 kPa	Solvent (water) and dissolved solutes	Suspended solids, fine particulars, some colloids
Ultrafiltration (UF)	Feed → / Retentate ↑ / Solvent ↑ / UF membrane	Pressure difference 100–800 kPa	Solvent (water) and low molecular weight solutes (<1000)	Macrosolutes and colloids
Nanofiltration (NF)	Feed → / Retentate ↑ / Permeate ↑ / NF membrane	Pressure difference 0.3–3 MPa	Solvent (water), low molecular weight solutes, monovalent ions	High molecular weight compounds (100–1000), multivalent ions
Reverse osmosis (RO)	Feed → / Retentate ↑ / Permeate ↑ / RO membrane	Pressure difference 1–10 MPa	Solvent (water)	Virtually all dissolved and suspended solids

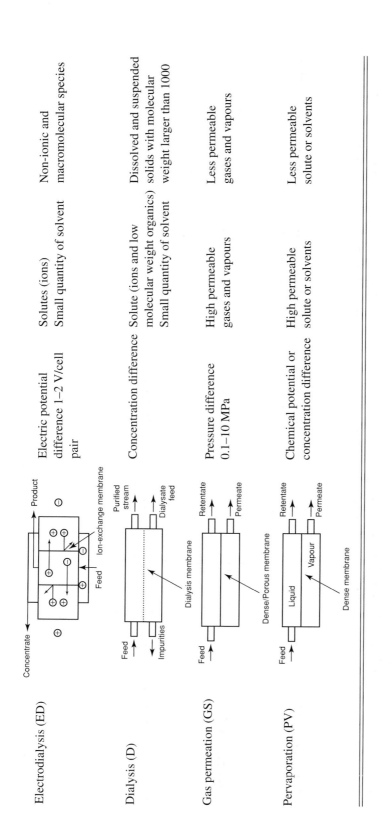

Process	Driving force	Permeate	Retained
Electrodialysis (ED)	Electric potential difference 1–2 V/cell pair	Solutes (ions) Small quantity of solvent	Non-ionic and macromolecular species
Dialysis (D)	Concentration difference	Solute (ions and low molecular weight organics) Small quantity of solvent	Dissolved and suspended solids with molecular weight larger than 1000
Gas permeation (GS)	Pressure difference 0.1–10 MPa	High permeable gases and vapours	Less permeable gases and vapours
Pervaporation (PV)	Chemical potential or concentration difference	High permeable solute or solvents	Less permeable solute or solvents

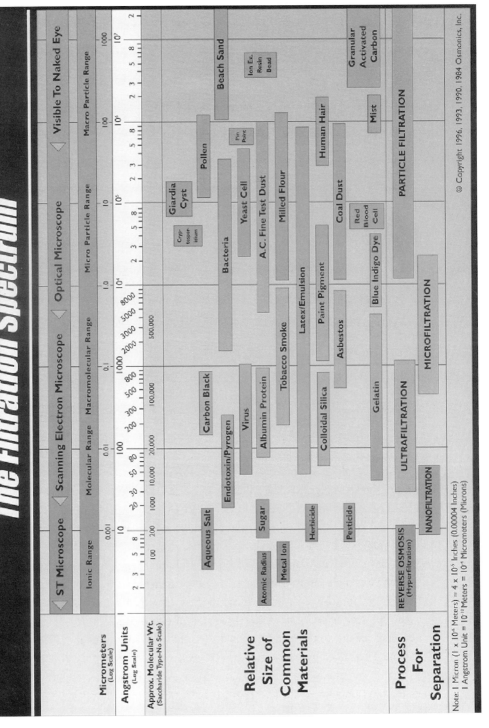

Figure 9.6

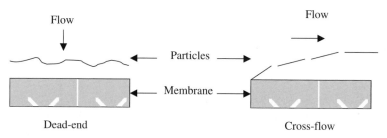

Figure 9.7 Basic filtration processes.

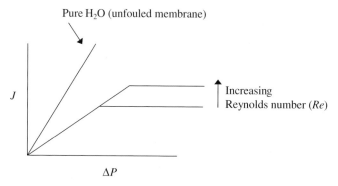

Figure 9.8 Solvent flux vs pressure drop.

The filtration process can operate in two basic modes, dead-end or cross-flow. These modes are shown in Figure 9.7.

In dead-end filtration, retained particles form a cake layer. The cake-layer thickness changes with time. As the thickness increases, the solvent flow decreases. In cross-flow operation, a large portion of the solvent passes parallel to the membrane without permeating through the membrane. The advantage of cross-flow operation is that the shear forces generated by the flow push the particles along the membrane, reducing fouling.

The solvent flux (J) is typically described by the equation $J = \Delta P / R_T$. The total resistance to flow (R_T) is expressed as the sum of two resistances, $R_m + R_c$, where R_m is the resistance due to the membrane and R_c is the cake-layer resistance. The resistance R_m can be determined by measuring the pure-water flux on an unfouled membrane, one limiting case corresponding to maximum solvent flux. This case is independent of feed flowrate. As R_c increases, the flux becomes independent of ΔP. This is illustrated in the Figure 9.8.

For a membrane that has been exposed to a feed solution, the solvent flux decreases compared to the pure-water value. As ΔP increases, the flux reaches a plateau value and becomes independent of ΔP. The plateau value is a function of the feed flowrate (shown by Re in Figure 9.8).

Example 9.2 [9]

Problem:

The petroleum industry and regulatory agencies are searching for alternative methods for treating drainwater from oil rigs. Membrane separation has proven to be an effective and economic alternative to conventional methods for oil–water separation. Drainwater from oil rigs may contain up to 5% oil in addition to different chemicals in varying concentrations. Drainwater typically is collected in a burner tank and burnt by flare to the atmosphere. This emission to air causes an unwanted environmental impact.

The membrane separation plant is tubular ultrafiltration (UF) and the pilot-plant operation was on a batch basis with a volume reduction factor approaching 40. The UF membrane had a maximum permeate flux of around 300 L/m^2 · hr at maximum 6 kg/cm^2 inlet pressure and 3.8 m/s fluid velocity with a clean membrane. The flux typically dropped and approached 80 L/m^2 · hr at the end of a day's operation. The retentate from UF separation was returned to the feed tank whereas the permeate was routed to the sewer. Design of a full-scale plant was performed using a flux value of 40 L/m^2 · hr and volume reduction of 20×.

Cleaning of the membrane was performed with 0.3 wt% nitric acid at 45 °C and the alkaline detergent Ultrasil 11 at 0.25 wt% and 55 °C. The membrane flux did not always fully recover after each chemical cleaning. A 200 mg/L Cl$_2$ solution administered as sodium hypochlorite (NaOCl) would, however, restore the original flux.

For the full-scale membrane plant, calculate:

(a) permeate and retentate flowrates;

(b) the membrane area required.

Solution:

Given: permeate flux = 40 L/m^2· hr;

        volume reduction of concentrate = 20 times.

<u>Mass balance on membrane unit</u> (refer to Figure 9.9):

(a) Permeate flowrate = x = 38 m^3/hr permeate × 10^3 L/m^3 = 3.8 × 10^4 L/hr;

    Retentate flowrate = 2 m^3/hr × 10^3 L/m^3 = 2 × 10^3 L/hr.

(b) Membrane area = $\dfrac{\text{permeate flowrate}}{\text{permeate flux}}$

$$= \frac{3.8 \times 10^4 \text{ L/hr}}{4.0 \times 10^1 \text{ L/m}^2 \cdot \text{hr}} = 950 \text{ m}^2.$$

[Note: assumes continuous operation.]

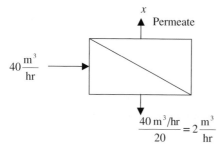

Figure 9.9 Schematic flow diagram, Example 9.2.

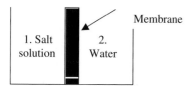

Figure 9.10 RO membrane.

9.9.4 Nanofiltration

Nanofiltration spans the gap in particle size between reverse osmosis (hyperfiltration) and ultrafiltration. It can separate high molecular weight compounds (100–1000) from solvents, and can also separate monovalent from multivalent ions. The driving force is a pressure difference of about 0.3–3 MPa (even greater than ultrafiltration). The nanofiltration process can reject selected (typically polyvalent) salts and may be used for selective removal of hardness ions in a process known as membrane softening [10].

9.9.5 Reverse osmosis

Osmosis refers to the flow of a pure solvent, usually water, across a solvent-permeable membrane. The flow is driven from the solvent phase towards a high salt concentration phase. The result is a dilution of the high salt concentration solution. Osmosis is not useful as a separation process since the solvent is moving in the wrong direction, resulting in dilution as opposed to separation. The solvent can be transferred in the opposite direction if a pressure is applied to the high salt concentration side of the membrane such that the pressure driving force is greater than the osmotic pressure gradient. Referring to Figure 9.10, the water flux is given by:

$$J_w = K(\Delta P - \Delta \Pi),\tag{9.16}$$

where K = permeability coefficient
$$\Delta P = P_1 - P_2$$
$$\Delta \Pi = \Pi_1 - \Pi_2$$
and $\Pi = CRT$;

Table 9.4 *Osmotic pressures.*

Compound	Concentration (g/L)	(Total moles in solution/L)	Osmotic pressure (atm, at 25 °C)
Sucrose	1	0.003	0.07
NaCl	58	2	49
NaCl	1	0.02	0.5
NaHCO$_3$	1	0.02	0.6
CaCl$_2$	1	0.03	0.7

C = total ionic concentration
R = universal gas constant
T = absolute temperature.

We can see from the flux equation that we have reverse osmosis if $\Delta P > \Delta \Pi$.

The osmotic pressure depends on the type of solution as well as the concentration in which it is present. Thus, 1 mole of NaCl dissolved in 1 L of water will double the osmotic pressure compared to 1 mole of glycerin added in the same amount of water, since the former yields two ions as opposed to only one molecule produced by glycerin. Similarly, 1 mole of ferric chloride (FeCl$_3$), by yielding four ions, will double the osmotic pressure of 1 mole of NaCl. Some typical osmotic pressures are given in Table 9.4.

As the pressure difference across the membrane is increased, the rate of solvent mass transfer is also increased. Typical feed pressures are between 17 and 55 atm (1.7–5.5 MPa). *Remember that the pressure difference must be greater than the osmotic pressure for reverse osmosis to occur, and that the osmotic pressure varies for different types of solutions and for the same solutions in different concentrations.*

As temperature varies, the diffusivity and viscosity vary also, and this in turn causes the flux to vary. Membrane area corrections $(A_T/A_{25\,°C})$ due to the respective temperatures are as follows [11]:

Temperature (°C):	10	15	20	25	30
Area correction:	1.58	1.34	1.15	1.00	0.84

Example 9.3: reverse osmosis

Problem:
A reverse osmosis unit is to demineralize 750,000 L/day treated effluent. Pertinent data are: permeability coefficient $= 0.2$ L/(m$^2 \cdot$ day $\cdot$ kPa) at 25 °C, pressure difference between the feed and product water $= 2500$ kPa, osmotic pressure difference between the feed and product water $= 300$ kPa, lowest operating temperature $= 10\,°$C.

Determine the membrane area required.

Solution:

The total water flux is given by:

$$N_w = 0.2 \text{ L/(m}^2 \cdot \text{day} \cdot \text{kPa)}(2500 \text{ kPa} - 300 \text{ kPa})$$
$$= 440 \text{ L/day} \cdot \text{m}^2 \text{ at } 25\,°\text{C}.$$

The membrane area at $10\,°\text{C}$ is given by:

$$A = (750{,}000 \text{ L/day})[(\text{day} \cdot \text{m}^2)/(440 \text{ L})](1.58) = 2690 \text{ m}^2.$$

Example 9.4

Problem:

Experiments at $25\,°\text{C}$ were performed to determine the water permeability and salt (NaCl) rejection of a cellulose acetate membrane. The membrane area is $A = 2.00 \times 10^{-3}$ m^2. The inlet salt concentration is $C_{s1} = 10.0$ kg NaCl/m^3 solution (10.0 g NaCl/L). The water recovery is low so that the salt concentration in the entering and exit feed solutions are assumed to be equal. The product solution contains $C_{s2} = 0.4$ kg NaCl/m^3 solution and the flowrate is 2×10^{-9} m^3 solution/s. A pressure differential of 55 atm is used. Calculate (a) the solute rejection R, and (b) the permeability coefficient of the membrane.

Solution:

The total salt flux across the membrane (N_s) is:

$$N_s = \frac{\left(2 \times 10^{-8}\,\frac{\text{m}^3 \text{ soln}}{\text{s}}\right)\left(\frac{0.4 \text{ kg salt}}{\text{m}^3 \text{ soln}}\right)}{2 \times 10^{-3} \text{ m}^2} = 4 \times 10^{-6}\,\frac{\text{kg salt}}{\text{m}^2 \cdot \text{s}}.$$

(a) The rejection R is:

$$R = \frac{C_{s1} - C_{s2}}{C_{s1}} = \frac{10 \text{ g/L} - 0.4 \text{ g/L}}{10 \text{ g/L}} = 0.96.$$

Thus, we can assume that the permeate flowrate is predominantly water. The total water flux (N_w) is:

$$N_w = \frac{\left(2 \times 10^{-8}\,\frac{\text{m}^3}{\text{s}}\right)\left(1 \times 10^3\,\frac{\text{kg H}_2\text{O}}{\text{m}^3}\right)}{2 \times 10^{-3} \text{ m}^2} = 1 \times 10^{-2}\,\frac{\text{kg H}_2\text{O}}{\text{m}^2 \cdot \text{s}}.$$

(b) To calculate the permeability coefficient, we first need to determine the osmotic pressure on each side of the membrane.

$$\Pi_1 = C_{s1}RT$$
$$= \left(10 \ \frac{g}{L}\right) \left(\frac{mol}{58 \ g}\right) \left(\frac{2 \ ions}{mol}\right) \left(82.05 \times 10^{-3} \ \frac{atm \cdot L}{mol \cdot K}\right) (298 \ K)$$
$$= 8.4 \ atm.$$

Similarly, $\Pi_2 = 0.3$ atm, so

$$K = \frac{N_w}{\Delta P - \Delta \Pi} = \frac{1.2 \times 10^{-2} \ \frac{kg \ H_2O}{m^2 \cdot s}}{55 \ atm - 8.1 \ atm} = 2.6 \times 10^{-4} \ \frac{kg \ H_2O}{m^2 \cdot s \cdot atm}.$$

Applications and characteristics

RO is not 100% effective, and some particles will pass through the membranes. The membranes will have a higher rejection efficiency for some particles than for others.

- Multivalent ions have a higher rejection than univalent ions (Ca^{2+} vs Na^+).
- Undissociated or poorly dissociated substances have lower rejection (e.g., silica).
- Acids and bases are rejected to a lesser extent than their salts.
- Co-ions affect the rejection of a particular ion (e.g., sodium is better rejected as sodium sulfate than as sodium chloride).
- Undissociated low-molecular-weight organic acids are poorly rejected and their salts are well rejected.
- Trace quantities of univalent ions are generally poorly rejected.
- The average rejection of nitrate is significantly below that of other common monovalent ions.

Hollow fibers appear to be the best in terms of surface area/volume, but they have serious hydraulic problems that cause the water flux to be much lower than other configurations.

The most common membrane is cellulose acetate. It can reject over 99% of salts, but the flux is relatively low: about 10 gal/ft$^2 \cdot$ day. It consists of a thin dense skin (2000 Å) on a porous support. The skin is the rejecting surface; the porous support is spongy and is $\frac{2}{3}$ water by weight. Another type of membrane is a polyamide membrane. It has a longer life than a cellulose-acetate membrane because it has a higher chemical and physical stability, and it is immune to biological attack. It is not prone to hydrolysis like cellulose acetate, and so it may be operated over a wider pH range. However, any water to be treated must be dechlorinated before contact with a polyamide membrane to avoid degradation.

Figure 9.11 illustrates generalized curves that show trends [12]. Figure 9.11(a) shows that the membrane flux increases linearly with pressure. The water quality (solute rejection) increases with pressure (Figure 9.11(b)) until concentration polarization builds up at the feed membrane surface and causes precipitation. Figure 9.11(c) and (d) show the effect of temperature on performance. As the temperature increases, the membrane material relaxes and water permeates faster. Along with this increased flux, there is an increase in the amount of salt that also permeates, reducing water quality. Figure 9.11(e) and (f) point out that performance decreases as the quantity of water recovered increases. This is

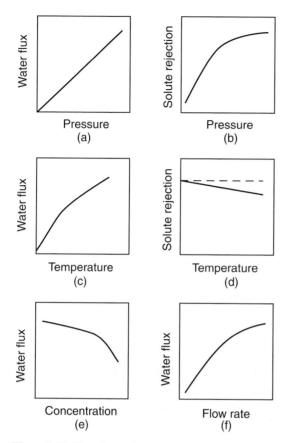

Figure 9.11 The effects of applied pressure, feed temperature, and water recovery on membrane flux and product water quality of RO membranes [12].

due to membrane compaction combined with an increase in concentration polarization. Concentration polarization will be described in detail later in the chapter.

RO is well-suited for use in treating secondary wastewater effluents, even though there will be some organic fouling. Cellulose-acetate membranes are especially useful for this purpose, since they can reject 90–99% of all salts and 90% of all organic material – all of this can be accomplished in one unit.

Some pilot plants are studying the feasibility of RO for demineralization of seawater and brackish water (Table 9.5), but the cost is very high. In seawater demineralization, $\frac{1}{2}$ of the entire cost is put towards replacement of the membranes. Other applications are: pretreatment of normal municipal water preceding ion exchange to make ultrapure water for applications such as boiler feed; recovery of valuable or reusable materials from a waste via the RO reject stream and reduction in the volume of waste, if required; and water conservation or recovery such as the cooling tower blowdown.

The flux value of the membranes used in RO will decrease over their lifetime as the pore passages decrease. This is a permanent and irreversible process, and the flux usually exhibits an exponential decay. However, it is possible to maximize the useful lifetime

Table 9.5 *Typical NF and RO performance specifications [after 12].*

Manufacturer	A	B	C	D
Process	Seawater RO	Brackish RO	Brackish RO	NF
Configuration	Spiral wound	Spiral wound	Spiral wound	Spiral wound
Membrane polymer	Polyamide	Composite polyamide	Cellulose-acetate blend	Composite polyamide
Area	27.87 m^2 (300 ft^2)	37.16 m^2 (400 ft^2)	49 m^2 (528 ft^2)	37 m^2 (400 ft^2)
Minimum salt rejection	99.6%	99.0%	98%	70% NaCl, 95% MgSO$_4$
Permeate flowrate	19 m^3/day (5 kgal/day)	45.4 m^3/day (12 kgal/day)	39.7 m^3/day (10.5 kgal/day)	47 m^3/day (12.5 kgal/day)
Max. applied pressure	8.3 MPa (1200 lb/in^2)	4.16 MPa (600 lb/in^2)	4.14 MPa (600 lb/in^2)	1.7 MPa (250 lb/in^2)
Max. feed flowrate	Process dependent	284 L/min (75 gal/min)	137.8 L/min (36.4 gal/min)	265 L/min (70 gal/min)
Max. operating temperature	45°C (113°F)	45°C (113°F)	40°C (104°F)	35°C (95°F)
Feed water pH range	4–11	3–10	4–6	3–9
Max. feed water turbidity	1.0 NTU	1.0 NTU	1.0 NTU	1.0 NTU
Max. chlorine concentration	0	< 0.1 mg/L	1.0 mg/L	< 0.1 mg/L
Single element recovery	17%	15%	20%	15%
Max. pressure drop per element	Process dependent	69 kPa (10 lb/in^2)	104 kPa (15 lb/in^2)	Process dependent

Data from manufacturer's literature.

of the membranes by avoiding scaling and organic fouling. Scales usually consist of compounds like calcium carbonate, calcium sulfate, silica, calcium phosphate, strontium sulfate, barium sulfate, and calcium fluoride. Pretreatment can reduce the effects of scaling; usually adjustment of pH or a softening process prior to RO will help. Metal-oxide fouling is best avoided by filtration or aeration plus filtration prior to RO.

9.9.6   Dialysis

Dialysis is not used very often in environmental applications, but a brief discussion is useful to compare it to other types of membrane processes. Dialysis separates solutes of different ionic or molecular size in a solution. The driving force is a difference in solute concentration across the membrane. The smaller ions and molecules will pass through the membrane, but the bigger particles cannot make it through the pore openings.

On one side of the membrane is the solvent, and on the other is the solution to be separated. The particles will pass from the solution side to the solvent side, in the direction of decreasing solute concentration. In a batch dialysis process, the mass transfer of solute passing through the membrane at a given time is:

$$\frac{\mathrm{d}M}{\mathrm{d}t} = K A(\Delta C), \tag{9.17}$$

where $\mathrm{d}M/\mathrm{d}t =$ mass transferred per unit time
$\qquad K =$ mass transfer coefficient
$\qquad A =$ membrane area
$\qquad \Delta C =$ difference in concentration of the solute passing through the membrane.

ΔC will decrease with time in the case of a batch dialysis cell, which is not at steady state. If the same type of membrane were run in a continuous process where the flow of the solution is countercurrent to the solvent flow direction, ΔC would be constant. In most applications, many of the cells are pressed together to make a stack, and all the cells are run in parallel.

One real application of membrane dialysis is the recovery of sodium hydroxide from a textile mill. The percent recovery was reported to be between 87.3% and 94.6%. However, dialysis is limited to smaller flowrates since the mass transfer coefficient (K) is relatively small. See Refs. [11,13] for additional information.

9.9.7   Electrodialysis

Electrodialysis involves the use of a selectively permeable membrane, but the driving force is an electrical potential across the membrane. Electrodialysis is useful for separating inorganic electrolytes from a solution, and can therefore be used to produce freshwater from brackish water or seawater. Electrodialysis typically consists of many cells arranged side by side, in a stack. Figure 9.12 illustrates a two-cell stack.

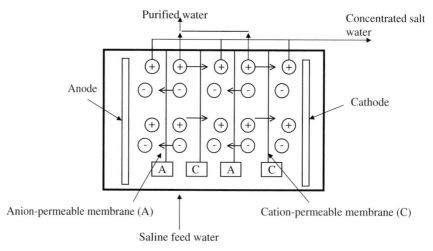

Figure 9.12 Two-cell electrodialysis stack.

When a constant voltage is applied to the electrodes, all cations migrate towards the cathode, and all anions migrate towards the anode. The cations can pass through the cation-permeable membrane, but they cannot permeate the anion-permeable membrane. The counter argument applies to anions. Alternate compartments contain ionic concentrations that are greater or less than the feed solution. These compartments are then combined to create the brine (waste) stream and the purified water stream.

The membranes are sheets fabricated of a synthetic ion-exchange resin. The cation-permeable membrane has a fixed negative charge (its fixed exchange sites are anionic). The cations in solution will enter the membrane when a voltage is applied to the system. They will not exchange with cations in the membrane because the electrical forces for ion motion are greater than the attractive forces between the cation and the membrane. Since the membrane structure is negatively charged, it repels anions. The opposite is true for the anion-permeable membrane.

The current required for an electrodialysis system can be calculated with Faraday's Law of electrolysis. One farad (F) (96,500 coulombs) will cause one gram equivalent weight of a charged species to migrate from one electrode to another:

Equivalents removed/unit time $= QNE_r$,

where Q = solution flowrate
N = normality of the solution (equivalents/L)
E_r = electrolyte removal (fraction of total equivalents).

The current for a single cell is:

$$I = \frac{FQNE_r}{E_c},$$

(9.18)

where I = current in ampères (A)

F = Faraday's constant (96,500 A · s/g equivalents removed)

E_c = current efficiency (fraction).

The current efficiency and the electrolyte removal are determined experimentally, and depend on the specific stack and feed conditions. The value of E_c is typically 0.90 or more, while E_r usually ranges from 0.25 to 0.50.

Since the same electrical current is passed through all of the cells in a stack, it is used n number of times (n = the number of cells). The current for the entire stack is:

$$I = \frac{FQNE_r}{nE_c}. \tag{9.19}$$

Two additional terms for an electrodialysis process are the current density and the normality of the feed water:

$$\text{Current density} = \frac{\text{current}}{\text{membrane area}}$$

$$\text{Feed water normality} = \frac{\text{equivalents}}{\text{volume of solution}}.$$

The current density to normality ratio describes the capacity of an electrodialysis cell to pass an electric current. This ratio may vary from 400 to 700 (when the current density is expressed as mA/cm^2). If the ratio is too high, regions of low ionic concentration will form near the membranes, which results in polarization. This causes high electrical resistance and a higher electrical consumption.

The power (W) required for an electrodialysis stack for treatment of a particular feed water is found from Ohm's Law:

$$V = RI \quad \text{and} \quad W = VI = RI^2.$$

Here, V is the voltage, and R is the resistance, which must be obtained experimentally and are specific for certain stack and feed water conditions.

Example 9.5: electrodialysis

Problem:

An electrodialysis stack having 100 cells is to be used to partially demineralize 100,000 L/day of wastewater. The salt content is 4000 mg/L and the cation or anion content is 0.066 gram equivalent weights per liter. Pilot-scale studies using a multicellular stack have been made. It was found that the current efficiency, E_c, was 0.9, the efficiency of salt removal, E_r, was 0.5, the resistance was 4.5 ohms, and the current density/normality ratio was 500. Determine:

(a) the current, I, required; (b) the area of the membranes; (c) the power required.

Solution:

(a) The current, I, is calculated for 100 cells:

$$I = \frac{FQNE_r}{nE_c}$$

$$= \frac{96{,}500 \text{ A} \cdot \text{s}}{\text{g} \cdot \text{eq} \cdot \text{wt}} \frac{100{,}000 \text{ L}}{\text{day}} \frac{0.066 \text{ g} \cdot \text{eq} \cdot \text{wt}}{\text{L}} \frac{1 \text{ day}}{86{,}400 \text{ s}} \frac{0.50}{100 \ 0.90}$$

$$= 41 \text{ A}.$$

Since the normality equals the number of gram equivalent weights per liter, normality $= 0.066$.

The current density is therefore equal to $(400)(\text{normality}) = (400)(0.066) = 33 \text{ mA/cm}^2$.

Thus, (b) the membrane area is:

$$\text{Area} = (41 \text{ A})(\text{cm}^2/33 \text{ mA})(1000 \text{ mA/A}) = 1240 \text{ cm}^2.$$

(c) The power required is:

$$\text{Power} = RI^2 = (4.5 \text{ ohms})(41 \text{ A})^2 = 7.6 \text{ kilowatts}.$$

Applications and characteristics

Electrodialysis is well-suited for demineralization of waters which contain 5 g/L or less of dissolved solids. Water can be purified to about 500 mg/L dissolved solids. The process is often too expensive to demineralize seawater, but there are full-scale plants in operation for the purpose of purifying brackish water.

Electrodialysis is also useful in secondary wastewater treatment. Since the membranes are susceptible to scaling and organic fouling, the wastewater must first be treated to reduce or eliminate scaling and fouling. Primary treatment can include coagulation, settling, acid addition, filtration or activated-carbon adsorption. The energy consumption of the process increases as scale deposits on the membrane (precipitated $CaCO_3$, for example). Membranes that have been fouled with organic compounds can be cleaned by treating them with an enzyme detergent solution. Pretreatment and cleaning is very important because about 40% of the total cost is for power consumption and membrane replacement. New developments are being made to try to reduce the complications involved with cleaning and replacing membranes [13]:

A recent innovation by a U.S. manufacturer of ED systems has significantly minimized the need for pretreatment. This innovation, known as electrodialysis reversal (EDR), operates on the same basic principle as the standard electrodialysis unit, except that both the product and the brine cells are identical in construction. At a frequency of 3–4 times/hr, the polarity of the electrodes is reversed and the flows are simultaneously switched by automatic valves, so that the product cell becomes the brine cell and the brine cell becomes the product cell. The salts are thus transferred in opposite directions across the membranes. Following the reversal of polarity and flow, the product water is discharged to waste until the cells and lines are flushed out and the desired water quality is restored. This process takes about 1–2 min. The reversal process aids in breaking up and flushing out scale, slimes, and other deposits in the cells. Scaling in the electrode compartments is minimized due to the continuous alternation of the environment from basic to acidic in the cells. Reversal of polarity, therefore, eliminates the need to continuously add acid and or SHMP. Some cleaning of the membrane stacks is still required, although with significantly reduced frequency than would be otherwise necessary.

9.9.8 Pervaporation

The word pervaporation is a contraction of two words, permeation and evaporation. This process is different from the other membrane processes in that there is a phase change as the solute permeates across the membrane. Thus, both heat and mass transfer are important aspects of the performance.

The feed is typically at atmospheric pressure and the permeate at high vacuum. As the solute permeates across the membrane, there is a change from liquid to vapor phase. Thus, both differences in permeation rates as well as heat of vaporization affect the productivity and selectivity of the process. Pervaporation offers the possibility of separating liquid solutions that are difficult to separate by distillation or other means.

The process is shown schematically in Figure 9.13. The feed liquid contacts one side of a membrane, which selectively permeates one of the feed components. The permeate, enriched in this component, is removed as a vapor from the opposite side of the membrane.

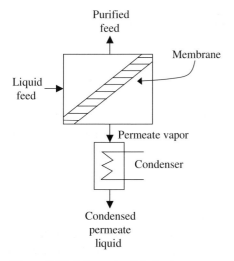

Figure 9.13 Schematic of the basic pervaporation process.

The driving force for the process is generated by the low pressure on the permeate side of the membrane, through cooling and condensing the permeate vapor.

Because of the phase change associated with the process and the non-ideal liquid-phase solutions (i.e., organic/water), the modeling of pervaporation cannot be accomplished using a solution–diffusion approach. Wijmans and Baker [14] express the driving force for permeation in terms of a vapor partial pressure difference. Because pressures on the both sides of the membrane are low, the gas phase follows the ideal gas law. The liquid on the feed side of the membrane is generally non-ideal.

The permeant flux of compound i is

$$N_i = \frac{Q}{\ell} \left(\gamma_i x_i P_i^s - y_i P_p \right) \tag{9.20}$$

where γ_i (activity coefficient) and x_i refer to the feed-side liquid, P_i^s is the vapor pressure at the feed-side temperature, y_i is the mole fraction in the permeant vapor, and P_p is the total permeant pressure.

The feed stream is usually heated to raise the equivalent vapor pressure on the feed side of the membrane. This results in a flux increase. The flux is also a strong function of permeate pressure. Maintaining a very low permeate pressure is important to maintain sufficient driving force.

The permeability for pervaporation depends on the concentrations of permeants in the polymer, which can cause swelling and solute-interaction effects in polymers. Inorganic membranes have recently been used in this application to overcome some of these limitations. Because of these non-ideal effects, the selectivity can be a strong function of feed concentration and permeate pressure, causing inversion of selectivity in some cases.

The separation selectivity can also be described by a separation factor, β_{pervap}, defined as:

$$\beta_{pervap} = \left(P_i^s / P_j^s \right)_{permeate} / \left(C_i^s / C_j^s \right)_{feed} \tag{9.21}$$

where C_i^s and C_j^s are the concentrations of components i and j on the feed (liquid) side and P_i^s and P_j^s are the vapor pressures of the two components i and j on the permeate (vapor) side of the membrane.

The separation factor β_{pervap} can be easily shown to be the product of two other separation factors, β_{mem} and β_{evap}, given by:

$$\beta_{mem} = \left(P_i^s / P_j^s \right)_{permeate} / \left(P_i^s / P_j^s \right)_{feed} \tag{9.22}$$

$$\beta_{evap} = \left(\frac{P_i^s / P_j^s}{C_i^s / C_j^s} \right)_{feed} . \tag{9.23}$$

It follows from equations above that:

$$\beta_{pervap} = \beta_{evap} \cdot \beta_{mem}. \tag{9.24}$$

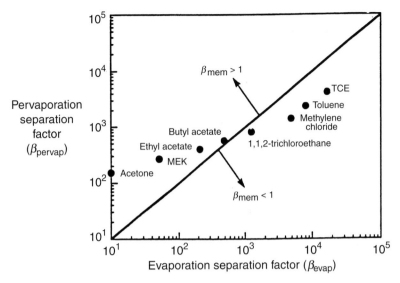

Figure 9.14 Pervaporation separation factor, β_{pervap}, as a function of the VOC evaporation separation factor, β_{evap}. Data obtained with laboratory-scale spiral-wound modules containing a silicone rubber membrane [14].

This equation shows that the separation is equal to the product of the separation due to evaporation of the liquid as well as separation by selective permeation through the membrane.

An example where β_{mem} is significant is the removal of water from an azeotropic mixture of water and alcohol. By definition, β_{evap} for an azeotropic mixture is 1. Thus, the 200- to 500-fold separation achieved is due entirely to the selectivity of the membrane.

In the separation of VOCs from dilute aqueous solutions, β_{evap} is often large and controlling relative to β_{mem}. Some data showing measured pervaporation separation factors for dilute aqueous VOC solutions are shown in Figure 9.14, in which β_{pervap} is plotted against β_{evap} obtained from Henry's Law data. Acetone has a pervaporation factor of approximately 150. In this case, β_{evap} is 10 and β_{mem} is 15. As the VOCs become more hydrophobic, the total separation factor increases, but the contribution of the membrane to the total separation decreases (data points relative to 45° line). For example, β_{pervap} of 500 for butyl acetate is almost all due to the evaporation step, since this point is very close to the 45° line. The total separation factor for TCE is 4,000, while β_{mem} is only 0.3. The decrease in membrane selectivity is unexpected because silicone rubber is hydrophobic. The selectivity for VOCs from water should increase as the VOCs become more hydrophobic. The reason given is concentration polarization.

Table 9.6 summarizes the relative ease of VOC separation using silicone rubber membranes in pervaporation mode. As the Henry's Law coefficient decreases, pervaporation becomes less selective and less competitive to other approaches, such as steam stripping.

Table 9.6 *Classes of VOCs based on volatility [14].*

Class	Henry's Law coefficient (atm/mole frac.)	Solubility in water (wt%)	Pervaporation separation factor with silicone rubber membranes	Ease of separation by pervaporation	Examples
Highly volatile	>200	<0.5	>1,000	Very good	TCE, toluene
Moderately volatile	0.5 to 200	0.5–infinite	20 to 1,000	Good to moderate	Methylene chloride, acetone, butanol
Non-volatile	<2	Infinite	<20	Poor	Ethanol, methanol

Example 9.6

Zeolite membranes are inorganic with uniform nanoporous structures. One important application is pervaporation to separate organic–water feed mixtures. A Ge-ZSM-5 membrane was used to separate 5 wt% organics in water mixtures. The results (Table 9.7) were obtained at 303 K with a membrane thickness $\ell = 30$ μm. The activity coefficient (γ) was calculated using the Wilson equation, Equation (3.6).

9.10 Factors that reduce membrane performance

The performance of membranes often decreases over time due to effects such as fouling and concentration polarization. This is seen as a decrease in flux (Figure 9.15). This performance decline is a major concern for filtration processes, but less so for gas separation processes.

Concentration polarization is a <u>reversible</u> increase in the concentration of retained solutes at the membrane feed interface. During operation, this effect will achieve a steady-state value and will disappear when the process is turned off.

Fouling is the <u>(ir)reversible</u> deposition of retained components on the membrane feed surface. These deposits can be biofilms, organic components, and/or inorganic salts. Accumulation tends to increase with time and the effect does not disappear when the process is turned off. There are some approaches to reducing the effect of fouling that will be described shortly. In Figure 9.16 we separate the flux decline in Figure 9.15 into these two effects. We can derive an equation to estimate the increase in retained solute concentration at the feed membrane surface due to concentration polarization.

Table 9.7 *Results of membrane performance, Example 9.6.*

Organic–water mixtures: feed

	wt% organic	x_i (mole frac.)	γ_i	P_i^{sat} (kPa)	Fugacity, $\hat{f}_{\text{org}}$ (kPa)
Ethanol	5	0.0202	6.19	10.39	1.30
Acetic acid	5	0.0155	2.79	2.77	0.12
Acetone	5	0.0161	11.64	38.04	7.13

Corresponding values for water

	wt% water	x_j (mole frac.)	γ_j	P_i^{sat} (kPa)	Fugacity, $\hat{f}_{\text{wat}}$ (kPa)
Ethanol	95	0.9798	1.006	4.23	4.17
Acetic acid	95	0.9845	1.004	4.23	4.18
Acetone	95	0.9839	1.004	4.23	4.18

Permeate $P_p = 0.5$ kPa

	Water flux, N_j (kg/m$^2 \cdot$ hr)	Organic flux, N_i (kg/m$^2 \cdot$ hr)	wt% water	y_j (water) (mole frac.)	y_i (org.) (mole frac.)
Ethanol	0.063	0.157	71.4	0.506	0.494
Acetic acid	0.066	0.0094	12.5	0.9589	0.0411
Acetone	0.035	0.289	89.2	0.281	0.719

Permeability calculated from Eqn (9.20)

	$Q\left(\dfrac{\text{kg} \cdot \text{m}}{\text{m}^2 \cdot \text{hr} \cdot \text{kPa}}\right)$ (organic)	$Q\left(\dfrac{\text{kg} \cdot \text{m}}{\text{m}^2 \cdot \text{hr} \cdot \text{kPa}}\right)$ (water)
Ethanol	1.24×10^{-12}	1.34×10^{-13}
Acetic acid	7.88×10^{-13}	1.49×10^{-13}
Acetone	3.56×10^{-13}	7.22×10^{-14}

Using Eqns (9.21)–(9.24)

	β_{pervap}	β_{evap}	β_{mem}
Ethanol	47	15.13	3.11
Acetic acid	2.71	1.76	1.54
Acetone	157	104.3	1.50

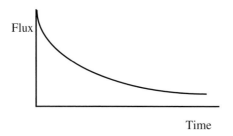

Figure 9.15 Flux behavior as a function of time.

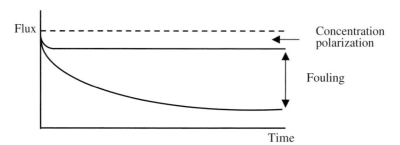

Figure 9.16 Flux as a function of time. Both concentration polarization and fouling are shown [2]. Reproduced with kind permission of Kluwer Academic Publishers.

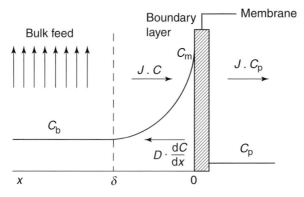

Figure 9.17 Concentration polarization; concentration profile under steady-state conditions [2]. Reproduced with kind permission of Kluwer Academic Publishers.

Using Figure 9.17: steady-state differential mass balance on the retained solute in the boundary layer is:

$$-v_p \frac{dC}{dx} = D \frac{d^2C}{dx^2}, \tag{9.25}$$

where v_p = permeate velocity.

Rearranging,

$$D\frac{d^2C}{dx^2} + v_p\frac{dC}{dx} = 0. \tag{9.26}$$

Integrating,

$$D\frac{dC}{dx} + v_pC = \text{constant} = v_pC_p \tag{9.27}$$
$$= \text{solute flux across the membrane.}$$

The boundary condition is:

$$x = 0; \quad C = C_m \tag{9.28}$$
$$= \text{membrane surface concentration.}$$

Solution for concentration profile is:

$$C = (C_m - C_p)e^{-(v_p/D)x} + C_p. \tag{9.29}$$

Rearranging,

$$\frac{C - C_p}{C_m - C_p} = e^{-(v_p/D)x}. \tag{9.30}$$

At $x = \delta$ (boundary-layer thickness), $C = C_b$ \hfill (9.31)

$$\frac{C_b - C_p}{C_m - C_p} = e^{-(v_p/D)\delta}. \tag{9.32}$$

Note:
- $D/\delta = k = $ mass transfer coefficient
- concentration profile is exponential
- when $v_p = 0$ (process off), effect disappears.

9.11 Effect of concentration polarization on membrane performance

$$R = \text{rejection coefficient} = 1 - \frac{C_p}{C_m} \tag{9.33}$$

$$R_{obs} = \text{observed rejection coefficient} = 1 - \frac{C_p}{C_b}. \tag{9.34}$$

We want to obtain an expression to relate the solvent flux (typically water) to measurable quantities. In general, we need to include the effect of osmotic pressure, especially for reverse osmosis applications.

$$J = K[\Delta P - \Delta \Pi] \tag{9.35}$$
$$= K[\Delta P - (\Pi_m - \Pi_p)] \quad \textit{Note}: \Pi_m \neq \Pi_b. \tag{9.36}$$

The van't Hoff equation is

$$\Pi = CRT \quad \text{(strictly valid for dilute solutions).} \tag{9.37}$$

Using equation (9.35)

$$\frac{\Pi_m}{\Pi_b} = \frac{C_m}{C_b}; \qquad \frac{\Pi_p}{\Pi_b} = \frac{C_p}{C_b}. \tag{9.38}$$

We can write the solvent flux as

$$J = K\left[\Delta P - \Pi_b\frac{C_m}{C_b} + \Pi_b\frac{C_p}{C_b}\right] \tag{9.39}$$

$$= K\left[\Delta P - \Pi_b\left(\frac{C_m}{C_b} - \frac{C_p}{C_b}\right)\right]. \tag{9.40}$$

Note:

$$\frac{C_m - C_p}{C_b - C_p} = e^{+(v_p/D)\delta} \tag{9.41}$$

$$\frac{C_m - C_p}{C_b} = \frac{C_m - C_p}{C_b - C_p} \cdot \frac{C_b - C_p}{C_b} \tag{9.42}$$

$$= e^{+(v_p/D)\delta}R_{obs}. \tag{9.43}$$

Substituting,

$$J = K\left[\Delta P - \Pi_b e^{(v_p/D)\delta}R_{obs}\right]. \tag{9.44}$$

Note: the solvent flux is a function of
- bulk feed osmotic pressure
- degree of concentration polarization
- observed rejection.

Specific correlations for concentration polarization using *flat sheet membranes* have been derived [15].

Laminar flow

$$\frac{C_m}{C_b} = 1.536(\varepsilon_o)^{1/3} + 1 \quad \varepsilon_o \ll 1. \tag{9.45}$$

Near channel entrance

$$\frac{C_m}{C_b} = \varepsilon_o + 5\left\{1 - \exp\left[-\left(\frac{\varepsilon_o}{3}\right)^{1/2}\right]\right\} + 1. \tag{9.46}$$

Far downstream

$$\frac{C_m}{C_b} = \frac{1}{3\sigma_o^2} + 1, \tag{9.47}$$

where

$$\sigma_0 = \frac{2D_{\text{solute}}}{v_p L} \qquad v_0 = \frac{v_p c}{3u_0 \sigma_0^2 h} \tag{9.48}$$

and h = half-width of channel

$\quad x$ = axial distance

$\varepsilon_0 = h/x$.

An important limit in concentration polarization is complete rejection ($R = 1$, $C_p = 0$). For organics, this can lead to gel layer formation which controls the permeation rate (v_p). For this limiting case, the equation for concentration polarization becomes

$$\frac{C_g}{C_b} = e^{+v_p/k} \tag{9.49}$$

C_g = gel layer concentration.

Rearranging and solving for v_p,

$$v_p = k \ln \left(\frac{C_g}{C_b} \right). \tag{9.50}$$

Mass transfer correlations have been derived that are used for membranes of various configurations [16]. For flow through *tubes*, the following are used (see Appendices A and B for details on dimensionless numbers and mass transfer correlations, respectively).

Turbulent flow

The Sherwood number (Sh) is given by:

$$Sh = 0.023 Re^{0.8} Sc^{0.33}. \tag{9.51}$$

Laminar flow

$$Sh = 1.86 \left(Re Sc \frac{d_h}{L} \right)^{0.33}, \tag{9.52}$$

where

$\quad d_h$ = hydraulic diameter

$$= \frac{4(\text{cross-sectional area})}{\text{wetted perimeter}} \tag{9.53}$$

and L = axial length of tube.

An alternative correlation for k uses the shear rate ($\dot{\gamma}$).

$$k = 0.816 \dot{\gamma}^{0.33} D^{0.67} L^{-0.33}, \tag{9.54}$$

where

$$\dot{\gamma} = \frac{8v}{d} \quad \text{(for tubes)} \tag{9.55}$$

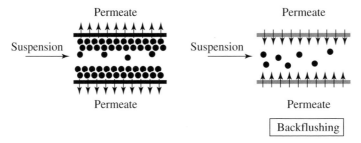

Figure 9.18 The principle of backflushing [2]. Reproduced with kind permission of Kluwer Academic Publishers.

and

$$\dot{\gamma} = \frac{6v}{b} \quad \text{(for rectangular channels)} \tag{9.56}$$

v = feed velocity
d = tube diameter
b = channel height.

Correlations are available in Appendix B to determine k for various configurations and operating conditions.

Pretreatment is often used to reduce fouling. Methods include heating, pH adjustment, chlorination, activated-carbon sorption, or chemical precipitation. Other factors such as membrane pore-size distribution, hydrophilicity/hydrophobicity, or surface charge can also reduce the effects of fouling. Methods which reduce concentration polarization, such as using higher axial flow velocities, lower flux membranes, or turbulence promoters, also help to reduce fouling.

Periodic changes in operation can help to reduce accumulation of foulants on the membrane surface. This can be accomplished with alternate pressurization and depressurization, changing the flow direction at a given frequency, or backflushing (Figure 9.18).

Backflushing can remove the fouling layer both at the membrane surface and within the membrane. The forward filtration time and the duration of the backpulse need to be optimized since permeate is lost to the feed side during the backpulse. A schematic is shown in Figure 9.19. See Ref. [17] for additional details.

Cleaning with chemical agents can be used. They need to be compatible with the membrane to avoid damaging the membrane structure. Some common cleaning agents are: acids (strong or weak), caustic (NaOH), detergents (alkaline, non-ionic), enzymes, complexers, and disinfectants.

9.12 Geomembranes

Some membrane applications are focused as barriers as opposed to separators. The clear film that covers meat products in the supermarket serve as barriers to reduce oxygen

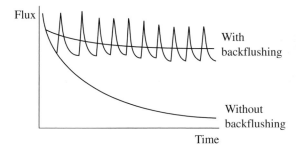

Figure 9.19 Schematic drawing of the flux vs time behavior in a given microfiltration process with and without backflushing [2]. Reproduced with kind permission of Kluwer Academic Publishers.

permeation and prolong shelf life. An environmental example is geomembranes. They are polymer sheets that are used as a liner for hazardous waste containment applications such as landfills. High-density polyethylene (HDPE) is the current choice for landfills containing a variety of waste products. HDPE is chemically compatible and resists degradation by a wide range of chemicals. Geomembranes need to be mechanically stable, resistant to degradation by a wide variety of mechanisms (chemical or biological attack, thermal swings, etc.) and serve as a barrier to both gas and liquid intrusion.

9.13 Remember

Membrane processes separate the components of a gas or a liquid stream by taking advantage of the fact that membrane materials are "selectively permeable." A membrane will allow certain species to pass, each at a different rate. There are various mechanisms that control this rate. Compared to most other processes, membrane separations use energy more efficiently, but have a relatively small mass flux.

Some important things to remember about membranes:

- Membrane processes are UNIT OPERATIONS. Regardless of what chemicals are being separated, the basic design principles for different types of membrane separations are always similar.
- Membrane processes are not reversible.
- Although membranes will usually separate a substance from a solution containing numerous substances with high efficiencies, fouling and scaling of the membranes can occur when insoluble species are encountered. Some membranes are also susceptible to biological attack.
- Membrane separations can occur in a batch or a continuous process. Most applications are continuous.
- Each membrane process requires a driving force for separation to occur. For example, dialysis requires a concentration difference, electrodialysis requires a difference in electric potential, and reverse osmosis requires a pressure difference.

9.14 Questions

9.1 What are the three important points with respect to the definition of a membrane?

9.2 What are three advantages and disadvantages of membrane processes?

9.3 Name two environmental applications for each membrane process discussed in this chapter.

9.4 Why is it that Knudsen diffusion typically cannot provide high selectivity for light gas separations?

9.5 What is the "upper bound" on a Robeson plot?

9.6 Is concentration polarization reversible?

9.15 Problems

9.1 Determine the osmotic pressure for a sodium-chloride solution with concentration = 23,400 mg/L at 25 °C.

9.2 A wastewater stream at 25 °C is to be treated using a reverse osmosis system. The wastewater has a flowrate of 50 gal/min and a concentration of 5000 mg/L of sodium chloride. A 75% recovery rate is required. A reverse osmosis vendor has recommended a membrane with the following characteristics:

- permeability coefficient = $1.9 \times 10^{-6} \dfrac{\text{gmol}}{\text{cm}^2 \cdot \text{s} \cdot \text{atm}}$
- area for a bundle = 300 ft^2
- 50% recovery rate
- optimal pressure differential across the membrane = 500 psi
- 95% rejection of the salt.

Determine:

(a) The osmotic pressure of the solution;

(b) the water flow through the membrane;

(c) the number of units required for 75% recovery.

9.3 A reverse-osmosis membrane to be used at 25 °C for a NaCl wastewater stream containing 2.5 g NaCl/L (2.5 kg NaCl/m^3, $\rho = 999$ kg/m^3) has a water permeability coefficient $K = 4.8 \times 10^{-4}$ kg/m$^2 \cdot$ s $\cdot$ atm and a solute (NaCl) permeability coefficient = 4.4×10^{-7} m/s. Calculate the water flux using a $\Delta P = 28$ atm and a solute rejection R. The permeate concentration is 0.1 kg NaCl/m^3.

9.4 An electroplating plant has 2,000 m^3/day of a nickel-bearing waste. The nickel concentration is 15,000 mg/L as NiSO$_4$. Assume the following characteristics of the system:

- resistance through unit = 10.5 ohms
- current efficiency = 90%
- maximum CD/N ratio = 6,000 (A/m^2)/(g-equivalent/L)
- membrane area = 1.0 m^2.

(a) Provide a preliminary design of a system to produce 90% removal of Ni.

(b) Determine the number of membranes,

(c) power required, and

(d) the annual electrical cost @ $0.05/kWh.

Hint: You can use the following equation to find the number of cells:

$$I = (FQN/n) \times (E_1/E_2),$$

where $I =$ current (amps)

   $F =$ Faraday' constant $= 96,487$ coulombs/g-equivalent (or amp · s/g-equivalent)

   $Q =$ flowrate (L/s) [gal/min $\times 0.06308 =$ L/s]

   $N =$ normality of solution (g-equivalents/L)

   $n =$ number of cells between electrodes

   $E_1 =$ removal efficiency (fraction)

   $E_2 =$ current efficiency (fraction)

Appendix A: Dimensionless numbers

A.1 Definition

A group of physical quantities with each quantity raised to a power such that all the units associated with the physical quantities cancel, i.e., dimensionless.

A.2 Buckingham Pi Theorem

This theorem provides a method to obtain the dimensionless groups which affect a process. First, it is important to obtain an understanding of the variables that can influence the process. Once you have this set of variables, you can use the Buckingham Pi Theorem. The theorem states that the number of dimensionless groups (designated as Π_i) is equal to the number (n) of *independent* variables minus the number (m) of dimensions. Once you obtain each Π, you can then write an expression:

$$\Pi_1 = f(\Pi_2, \Pi_3, \ldots).$$

This result only gives you the fact that Π_1 can be written as a function of the other Πs. Normally, the exact functional form comes from data correlation or rearrangement of analytical solutions. Correlating data using the dimensionless numbers formed by this method typically allows one to obtain graphical plots which are simpler to use and/or equations which fit to the data. If the dimensionless terms are properly grouped, they represent ratios of various effects and one can ascertain the relative importance of these effects for a given set of conditions.

Basic dimensions include length (L), time (t), and mass (M). Variables which are independent means that you cannot form one variable from a combination of the other variables

Table A.1

Variable	Dimensions
k = mass transfer coefficient	L/t
v_∞ = free stream fluid velocity	L/t
v = kinematic viscosity of fluid	L^2/t
D_{AB} = solute diffusion coefficient	L^2/t
L = plate length	L

in the set. As an example, with density (ρ), viscosity (μ) and kinematic viscosity (v), only two of the three variables are independent. Can you see why?

As an example of the method, let us examine the situation of mass transfer associated with flow across a flat plate. The variables which are important are listed in Table A.1, together with their dimensions. In this case, $\Pi = n - m = 5 - 2 = 3$.

Choose two variables, say L and v, which do not form a dimensionless set. This leaves three variables for use to form the three Πs. We then combine L and v together with each of the other three variables so that each result is dimensionless. The method is to raise each variable to the appropriate power so that each dimension sums to zero. Choosing k as the variable:

$$L^a v^b k = L^a L^{2b} t^{-b} L t^{-1}.$$

We can write an equation for the sum of the exponents for each dimension. Each equation must sum to zero for the resulting group to be dimensionless.

$$\left. \begin{array}{ll} \text{L:}\ a + 2b + 1 = 0 & a = 1 \\ \text{t:}\ -b - 1 = 0 & b = -1 \end{array} \right\} \Pi_1 = \left\{ \frac{kL}{v} \right\}.$$

Similarly, using D_{AB} as the other variable gives: $\Pi_2 = \dfrac{D_{AB}}{v}$, and using v_∞ gives: $\Pi_3 = \dfrac{v_\infty L}{v}$. We can then write:

$$\frac{kL}{v} = f\left(\frac{v_\infty L}{v}, \frac{v}{D_{AB}} \right).$$

We can recognize two dimensionless groups as:

$$Re = \frac{v_\infty L}{v}; \quad Sc = \frac{v}{D_{AB}},$$

where Re and Sc are known as the Reynolds and the Schmidt numbers, respectively. The group $\dfrac{kL}{v}$ is usually not used, $\dfrac{kL}{D_{AB}} = Sh$ being a better choice (Sh is the Sherwood number). It is equivalent to use either since we would still have three Πs. We have not violated any constraints with this rearrangement. We have simply multiplied both sides of our original functional form (kL/ν) by Sc.

This illustrates an important point. With this method, we can obtain dimensionless numbers but they may not be in the best (or most recognizable) form. So the final form that is normally used is:

$$Sh = f(Re, Sc).$$

These dimensionless numbers have physical significance which is usually more useful than the original set when correlating data.

Correlations for mass transfer coefficients in this and other texts will have this functional form for similar flow situations.

A.3 Putting an equation in dimensionless form

Consider flow across a horizontal flat plate.

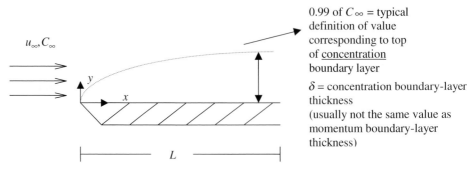

0.99 of C_∞ = typical definition of value corresponding to top of <u>concentration</u> boundary layer

δ = concentration boundary-layer thickness (usually not the same value as momentum boundary-layer thickness)

u: velocity in x-direction
v: velocity in y-direction

• Steady-state convective diffusion equation (x-direction) after simplification:

$$\left[u\frac{\partial C_A}{\partial x} + v\frac{\partial C_A}{\partial y} \right] = D_{AB}\frac{\partial^2 C_A}{\partial y^2}.$$

Each <u>variable</u> needs to be incorporated into a dimensionless form.
Each constant (i.e., D_{AB}) is left "as is."

Note: When putting variables in dimensionless terms, use quantities that can be measured or determined (i.e., L, C_∞, u_∞). Note that δ is not a good choice.
- Select dimensionless variables:

$$x^* = x/L; \qquad y^* = y/L,$$

(note: L used here and not δ since δ is a variable and unknown)

$$u^* = u/u_\infty; \qquad v^* = v/u_\infty,$$

(note: u_∞ used here since there is no constant value of v in the y-direction)

$$C_A^* = C_A/C_\infty.$$

Rearrange the dimensionless variables and substitute into the equation:

$$x = x^*L, \quad y = y^*L, \quad u = u^*u_\infty, \quad v = v^*u_\infty, \quad C_A = C_A^*C_\infty,$$

$$u^*u_\infty \frac{C_\infty}{L} \frac{\partial C_A^*}{\partial x^*} + v^*u_\infty \frac{C_\infty}{L} \frac{\partial C_A^*}{\partial y^*} = \frac{D_{AB}C_\infty}{L^2} \frac{\partial^2 C_A^*}{\partial y^{*2}}.$$

Rearrange:

$$u^* \frac{\partial C_A^*}{\partial x^*} + v^* \frac{\partial C_A^*}{\partial y^*} = \frac{D_{AB}}{Lu_\infty} \frac{\partial^2 C_A^*}{\partial y^{*2}} = \left(\frac{D_{AB}}{\nu}\right) \left(\frac{\nu}{Lu_\infty}\right) \frac{\partial^2 C_A^*}{\partial y^{*2}} = \frac{1}{Sc\,Re} \frac{\partial^2 C_A^*}{\partial y^{*2}}.$$

If $Re Sc$ is large (inertial effects dominate), one can neglect right side of equation.
If $Re Sc$ is small (diffusional effects dominate), right hand side of equation is significant.

A.4 Summary of the uses of dimensionless numbers

A.4.1 Correlate data

This is one important use of the Buckingham Pi Theorem. When the relationship between the various quantities is unknown (i.e., there is no equation to relate them), the dimensionless numbers provide a basis for obtaining an equation which fits the data. The exact coefficients and exponents for the dimensionless numbers are obtained from a best fit of the experimental data.

Example

$$Sh = 0.626\,Re^{1/2}Sc^{1/3}$$

(correlates the average value of k for flow over a flat plate of length L).

A.4.2 Scale

A dimensionless number can provide an estimate of the relative effect of various phenomena.

Examples

$$Re = \frac{\text{inertial forces}}{\text{viscous forces}}.$$

A large value of Re indicates that inertial forces dominate. Conversely, a small value means that viscous forces are more important.

$$Sh = \frac{kL}{D_{AB}} = \frac{\text{mass transfer resistance in material adjacent to fluid phase}}{\text{mass transfer resistance in fluid boundary layer}}.$$

A large value of Sh indicates that the mass transfer resistance in the fluid boundary layer is insignificant. A small value means that boundary layer mass transfer resistance dominates.

$$Sc = \frac{\nu}{D_{AB}} = \frac{\text{fluid kinematic viscosity}}{\text{solute diffusion coefficient in fluid phase}}$$
$$= \frac{\text{momentum diffusivity}}{\text{mass diffusivity}}.$$

The relative size of the momentum (M) and concentration (C) boundary layers is given by:

$$\frac{\delta_M}{\delta_C} = Sc^n.$$

A dimensionless number can also indicate the relative importance of terms in an equation. See the Re example given above.

A.4.3 Solve an equation for a variety of values of the physical quantities

The type of equation and initial or boundary conditions do not change, only their values.

Example

$$\frac{dC}{dt} = -\frac{kA}{V}(C - C_\infty).$$

Initial conditions: $t = 0; \quad C = C_i.$

This equation describes the concentration change in a batch vessel of volume V which contains a sorbent with total surface area A. The concentration C_∞ is that which corresponds to the maximum loading of the sorbent. Mass transfer to the particle surface controls the sorption. The solution can be rearranged to give:

$$\theta^* = \frac{C - C_\infty}{C_i - C_\infty} = e^{-\tau}; \quad \tau = \frac{kAt}{V}.$$

The solution in this form shows that θ^* is only a function of τ. Therefore, we can make one graph for all solutions:

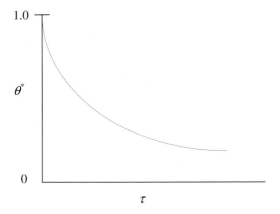

Check for yourself that you could change the values of k, A, V, C_i, or C_∞ and the above graph would still be valid.

Appendix B: Mass transfer coefficient correlations

B.1 Definition(s)

There are several common definitions for a mass transfer coefficient, k. These definitions are based on the driving force used. The typical units are different which is a "flag" as to the proper equation to use.

Equation	Typical units for k
$N_A = k\left(C_{A_1} - C_{A_2}\right)$	cm/s
$N_A = k\left(P_{A_1} - P_{A_2}\right)$	$\mathrm{mol/cm^2 \cdot s \cdot atm}$
$N_A = k\left(x_{A_1} - x_{A_2}\right)$	$\mathrm{mol/cm^2 \cdot s}$

There are two generic sources for obtaining mass transfer coefficients. First, you can find a correlation directly for a mass transfer coefficient. Second, you can find a correlation for a heat transfer coefficient in the analogous circumstances. For the former, you can proceed directly to calculate the mass transfer coefficient. In the latter case, you first need to substitute the appropriate dimensionless numbers to convert the correlation from the heat transfer to the mass transfer case.

Heat transfer	Mass transfer
k = heat transfer coefficient	k = mass transfer coefficient
L = characteristic length	L = characteristic length
$Nu = \dfrac{hL}{k}$	$Sh = \dfrac{kL}{D_{AB}}$
$Pr = \dfrac{\nu}{\alpha}$	$Sc = \dfrac{\nu}{D_{AB}}$

$$St = \frac{Nu}{Re\,Pr} \qquad St = \frac{Sh}{Re\,Sc}$$

$$Pe = Re\,Pr \qquad Pe = Re\,Sc$$

(dimensionless numbers are: Nu = Nusselt; Pr = Prandlt; St = Stanton; Re = Reynolds; Pe = Peclet; Sh = Sherwood; Sc = Schmidt).

Example

In forced convection around a sphere of diameter D, the correlation for the heat transfer coefficient associated with heat transfer between the bulk fluid and the sphere surface is:

$$Nu = 2 + Re^{1/2}\,Pr^{1/3}.$$

For the <u>same</u> physical situation, the mass transfer correlation would be:

$$Sh = 2 + Re^{1/2}\,Sc^{1/3}.$$

There are two important points to mention at this time.

1 The characteristic length that is used must be given with the correlation. It can be a linear length, a radius, a diameter, etc. So, be certain.

2 The dimensions associated with the mass transfer coefficient can vary depending on the flux equation used to define k. So, again, be certain, since the dimensions can change the functional form of the correlation and you want to make sure that the k you calculate is the one that you need.

B.2 Methodology for selecting the proper correlation for a mass transfer coefficient

1 What is the exact nature of the fluid flow? As an example, take a hollow tube (pipe) as the solid surface. Is the flow inside or outside the tube? Is the tube horizontal or vertical? Is there just one tube or are there several? If there are several, what is the pattern of their arrangement? You can also have fluid–fluid interactions (gas–liquid, for example).

2 For dilute binary mixtures, use the fluid properties of the concentrated component. This is reasonable since the concentration of the dilute component will have a negligible effect on the physical properties of the fluid. For concentrated solution, you will need to use averaging rules for that property.

3 Calculate the Reynolds number (Re). This value will indicate if the flow is laminar or turbulent. Be careful; the transition value of Re depends on the flow configuration.

4 Determine the type of coefficient desired. Do you want a local value (at a point on the solid surface) or an average value over a length?

You can now check various correlations to find the one that matches your needs. Most heat transfer or mass transfer texts contain additional correlations.

B.3 Mass transfer correlations

[Excerpts from: *Diffusion: Mass Transfer in Fluid Systems*, E.L. Cussler, Cambridge University Press, 1997]

B.3.1 Fluid–solid and fluid–fluid interactions (see Table B.1)

B.3.2 Fluid–column packings

These correlations are based on data for the following column packings:
Raschig rings: $\frac{1}{4}$ to 2 inches
Berl saddles: $\frac{1}{2}$ to $1\frac{1}{2}$ inches
Pall rings: 1 inch
Spheres: $\frac{1}{2}$ and 1 inch
Rods: $\frac{1}{2}$ and 1 inch

Liquid phase:
$$k_L \left(\frac{\rho_L}{\mu_L g} \right)^{1/3} = 0.0051 \left(\frac{G_L}{a_w \mu_L} \right)^{2/3} \left(\frac{\mu_L}{\rho_L D_L} \right)^{-1/2} (a_v D_p)^{0.4}.$$

Gas phase:
$$\frac{k_G RT}{a_v D_G} = C_1 \left(\frac{G_G}{a_v \mu_G} \right)^{0.7} \left(\frac{\mu_G}{\rho_G D_G} \right)^{1/3} (a_v D_p)^{-2.0}.$$

The range of data for k_L extended from a liquid Reynolds number $(G_L/a_w \mu_L)$ of 4 to 400 for water and 1 to 40 for organic absorbents. For k_G, the gas Reynolds number $(G_G/a_v \mu_G)$ ranged from 5 to 1000. For the packing:

$$\frac{a_{\text{wet}}}{a_v} = 1 - \exp \left[-1.45 \left(\frac{\sigma_c}{\sigma} \right)^{0.75} \left(\frac{G_L}{a_v \mu_L} \right)^{0.1} \times \left(\frac{G_L^2 a_v}{\rho_L^2 g} \right)^{-0.05} \left(\frac{G_L^2}{\rho_L \sigma a_v} \right)^{0.2} \right],$$

where a_v = specific surface area of packing, (m^2/m^3)
a_{wet} = specific surface area of wetted packing, (m^2/m^3)
C_1 = dimensionless constant. $C_1 = 5.23$ for packing larger than $\frac{1}{2}$ inch;
    $C_1 = 2.0$ for packing less than $\frac{1}{2}$ inch
D_L = liquid-phase diffusion coefficient, (m^2/s)
D_p = nominal packing size, (m)
D_G = gas-phase diffusion coefficient, (m^2/s)
g = acceleration of gravity, (9.81 m/s^2)
G_L = liquid-phase mass velocity, $(kg/s \cdot m^2)$
G_G = gas-phase mass velocity, $(kg/s \cdot m^2)$
k_G = individual gas-phase mass transfer coefficient, $(mol/s \cdot m^2)$
k_L = individual liquid-phase mass transfer coefficient, $(mol/s \cdot m^2)$
μ_L = liquid viscosity, $(kg/m \cdot s)$
μ_G = gas viscosity, $(kg/m \cdot s)$

Table B.1 *Fluid–solid and fluid–fluid interaction correlations.*

Interaction	Correlation	Comment
Fluid–solid		
• Flow through circular pipe		Characteristic length = pipe diameter
(a) Laminar	$Sh = 1.86(ReSc)^{0.8}$	Velocity = average fluid velocity in
(b) Turbulent	$Sh = 0.026Re^{0.8}Sc^{1/3}$	pipe
• Flow through non-circular cross-sections		Use hydraulic radius to determine equivalent diameter. Then use above correlations
• Forced convection around a solid sphere	$Sh = 2 + 0.6Re^{1/2}Sc^{1/3}$	Characteristic length = sphere diameter Velocity = relative velocity of sphere and bulk fluid
• Laminar flow across a flat plate		Characteristic length = plate length
(a) Local value	$Sh = 0.322Re^{1/2}Sc^{1/3}$	
(b) Average value	$Sh = 0.664Re^{1/2}Sc^{1/3}$	
• Spinning disc	$Sh = 0.62Re^{1/2}Sc^{1/3}$	$Re = \dfrac{D^2\omega}{v}$ D = disc diameter ω = disc rotation rate
• Packed beds of spherical particles	$k/v = 1.17Re^{-0.42}Sc^{-0.67}$	Characteristic length = particle diameter v = superficial velocity
Fluid–fluid		
• Dispersed phase (drops or bubbles) in stirred continuous phase	$Sh = 0.13\left[\dfrac{L^4(W/V)}{\rho v^3}\right]Sc^{1/3}$	L = stirrer length = characteristic length W/V = power per total fluid volume

ρ_L = liquid density, (kg/m^3)

ρ_G = gas density, (kg/m^3)

σ = liquid surface tension, (dyne/cm)

σ_c = critical surface tension of packing material, (dyne/cm)

Values of σ_c for selected materials are given in Table B.2.

Table B.2 *Critical surface tension of packing materials.*

Material	σ_c, dyne/cm
Carbon	56
Ceramic	61
Glass	73
Paraffin	20
Polyethylene	33
Polyvinylchloride	40
Steel	75

Source: A. S. Foust, L. A. Wenzel, C. W. Clump, L. Maus, L. B. Anderson, *Principles of Unit Operations,* 2nd edn (New York: John Wiley and Sons, 1980). This material is used by permission of J. Wiley & Sons, Inc.

Appendix C: Pulse analysis

Pulse analysis is a means to couple experimental measurements with a mass transfer model of the system to evaluate various parameters in the model. The experimental measurements are straightforward. A pulse, typically a square wave, of a solute enters the inlet of the system. The concentration profile of the solute at the system outlet is measured. This is shown graphically in Figure C.1. The mass transfer model is solved for the solute concentration using Laplace transforms. The solute concentration and various derivatives in the Laplace domain will be shown to be related to various moments of the concentration vs time data.

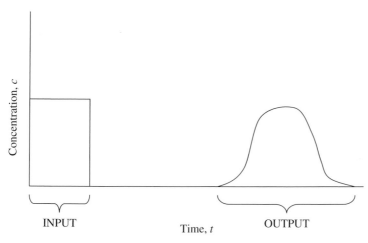

Figure C.1

C.1 Moments

We now have solute concentration vs time data for both the input and output profiles. The data can be integrated to calculate various moments as defined below.

C.1.1 Absolute moments

$$m_0 = \int_0^\infty c(t)\,dt = \text{constant if system doesn't lose mass}$$

$$m_1 = \int_0^\infty t \cdot c(t)\,dt$$

$$m_2 = \int_0^\infty t^2 \cdot c(t)\,dt.$$

We can normalize these values by dividing by m_0 $(z = 0)$

$$\mu_1' = \frac{m_1}{m_0} = \text{center of gravity of the profile}$$

$$\mu_2' = \frac{m_2}{m_0}.$$

C.1.2 Centralized moments (normalized)

We can also define moments where the time axis is shifted to the center of gravity.

$$\mu_2 = \frac{1}{m_0} \int_0^\infty (t - \mu_1')^2 c(t)\,dt \Rightarrow \text{measure of width of curve}$$

$$\mu_3 = \frac{1}{m_0} \int_0^\infty (t - \mu_1')^3 c(t)\,dt \Rightarrow \text{measure of asymmetry}$$

$$\mu_1 = \frac{1}{m_0} \int_0^\infty (t - \mu_1')c(t)\,dt = \mu_1' - \mu_1' \underbrace{\left[\frac{1}{m_0} \int_0^\infty c(t)\,dt \right]}_{m_0} = 0.$$

We now introduce the use of Laplace transforms to illustrate how we can use the solute concentration and various derivatives in the Laplace domain to obtain equations for the various moments.

C.2 Laplace transforms

$$\bar{c} = \int_0^\infty e^{-st} c(t)\, dt \quad \text{Definition of Laplace transform}$$

$$\frac{d\bar{c}}{ds} = \int_0^\infty (-t)e^{-st} c(t)\, dt.$$

We can generalize this to the kth derivative. If we take the limit as $s \to 0$, we see that we obtain the equation for the kth moment multiplied by $(-1)^k$:

$$\lim_{s \to 0} \left(\frac{d^k \bar{c}}{ds^k} \right) = (-1)^k \int_0^\infty t^k c(t)\, dt = (-1)^k m_k.$$

The approach is to obtain the experimental measurements and calculate numerical values for the various moments. The model equation is solved in the Laplace domain to obtain equations for the various moments. Equating the moment equations to the numerical values allows one to solve for various unknowns in the model equation(s).

In theory, we could generate several moment equations and solve for an equivalent number of parameters. In practice, it is best to only use the first and second moments. The outlet concentration profile usually has some "tailing" at longer times. These values usually have the largest error in their values and can skew the calculated moment values since the concentration is multiplied by t^k for the kth moment.

C.2.1 Example C.1: inert packed bed (see Figure C.2)

Problem:
Use the Laplace transform to model mass transfer in a cylindrical packed bed.

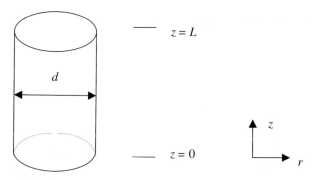

Figure C.2

Assumptions:

1 Neglect porosity of packing (solid particles)
2 No chemical reactions in the packed bed
3 Neglect changes in radial direction.

The differential mass balance on the fluid phase is:

$$\frac{\partial c}{\partial t} = -\vec{\nabla} \cdot \vec{N}_z + R_i^0$$

where

$$N_z = cv - D\frac{\partial c}{\partial z}$$

$$cv = \text{convective transport}$$

$$-D\frac{\partial c}{\partial z} = \text{axial dispersion.}$$

Open tube (no packing):

$$\frac{\partial c}{\partial t} = -V_0\frac{\partial c}{\partial z} + D_0\frac{\partial^2 c}{\partial z^2},$$

where D_0 = axial dispersion coefficient
V_0 = superficial velocity (Q/A)
and Q = volumetric flowrate
A = cross-sectional area.

Packed tube:

$$\frac{\partial c}{\partial t} = -v\frac{\partial c}{\partial z} + D\frac{\partial^2 c}{\partial z^2}$$

$$v = \frac{v_0}{\alpha}$$

$$D = \frac{D_0}{\alpha}$$

$$\alpha = \text{void fraction.}$$

Rearranging, we obtain the model equation:

$$\alpha\frac{\partial c}{\partial t} = -V_0\frac{\partial c}{\partial z} + D_0\frac{\partial^2 c}{\partial z^2}$$

I.C. $c(z \geq 0, t \leq 0) = 0$

B.C. $c(z = 0, t > 0) = c_0(t)$

$c(z \rightarrow \infty, t > 0) = \text{finite.}$

The initial condition (I.C.) states that no solute is initially in the system. The boundary condition (B.C.) at $z = 0$ is the input pulse. The second boundary condition means that the solution must remain finite for any axial length of the bed.

Take the Laplace transform:

$$\frac{D_0}{\alpha}\frac{d^2\bar{c}}{dz^2} - \frac{V_0}{\alpha}\frac{d\bar{c}}{dz} - s\vec{\bar{c}} + c(z, t = 0) = 0$$

where

$$\bar{c}(z = 0, s) = \bar{c}_0(s)$$
$$\bar{c}(z \to \infty, s) = \text{finite}$$
$$\text{Assume} \quad \bar{c} = k e^{\lambda z}.$$

The characteristic equation is:

$$\frac{D_0}{\alpha}\lambda^2 - \frac{V_0}{\alpha}\lambda - s = 0$$

$$\lambda = \frac{\alpha}{2D_0}\left[\frac{V_0}{\alpha} \pm \sqrt{\left(\frac{V_0}{\alpha}\right)^2 + 4s\frac{D_0}{\alpha}}\right] \qquad \text{use only negative root}$$

$$k = \bar{c}_0(s).$$

The solution for $\bar{c}$ is:

$$\bar{c}(z, s) = \bar{c}_0(s)e^{\lambda z},$$

where λ is defined above. We can now obtain an equation for various moments:

$$m_0(z) = \lim_{s \to 0} \bar{c}_0(s)e^{\lambda z} = \bar{c}_0$$
$$m_1(z) = -\lim_{s \to 0}\frac{d\bar{c}}{ds}.$$

To calculate $m_1(z)$

$$\frac{d\bar{c}}{ds} = \frac{d\bar{c}_0}{ds}e^{\lambda z} + \bar{c}_0(s)e^{\lambda z}\frac{V_0 z(-1/2)4\alpha\left(D_0/V_0^2\right)}{2D_0\sqrt{1 + 4s\alpha\left(D_0/V_0^2\right)}}$$

$$\underbrace{\lim_{s \to 0}\frac{d\bar{c}}{ds}}_{-m_1(z)} = \underbrace{\lim_{s \to 0}\left\{\frac{d\bar{c}_0}{ds}\cancel{e^{\lambda z}}^{1}\right.}_{-m_1(z = 0)} + \left. m_0\cancel{e^{\lambda z}}^{1}\left[-\frac{\alpha z}{V_0}\right]\right\}.$$

Rearranging, and using the definitions for the normalized moment:

$$\mu_1'(z) = \mu_1'(z = 0) + \frac{\alpha z}{V_0}$$

$$\Delta\mu_1' = \mu_1'(L) - \mu_1'(z = 0) = \frac{\alpha L}{V_0} = \frac{L}{v}$$

$$= \text{time for pulse to travel from } z = 0 \text{ to } z = L \text{ (time of travel)}$$

$$v = \frac{L}{\Delta\mu_1'} = \frac{V_0}{\alpha}$$

$$V_0 = \frac{\alpha L}{\Delta\mu_1'} = \text{average superficial fluid velocity.}$$

It can be shown that:

$$\mu_2 = \mu_2(z=0) + \frac{2zD_0\alpha^2}{V_0^3}$$

$$\Delta\mu_2 = \mu_2(L) - \mu_2(0) = \frac{2LD_0\alpha^2}{V_0^3} = \frac{2LD_0}{v\alpha}.$$

Application of the method

Packed tube:

1 Measure V_0.

2 Perform pulse injection experiment.

3 Calculate two system parameters using moments calculated from experimental results:

$$\alpha = \frac{V_0}{L}(\Delta\mu_1')$$

$$D_0 = \frac{V_0^3}{2L\alpha^2}(\Delta\mu_2).$$

Open tube ($\alpha = 1$):

$$V_0 = \frac{L}{\Delta\mu_1'}$$

$$D_0 = \frac{V_0^3}{2L}(\Delta\mu_2).$$

Numerical values of $\Delta\mu_1'$ and $\Delta\mu_2$ are obtained from integration of the experimental measurements using the definition of the first and second moments.

[Note: all results to this point are valid *only* for *inert* packed beds.]

C.2.2 Example C.2

Consider a packed column where we need to account for:

1 pore diffusion;

2 mass transfer to particle surface;

3 adsorption at the pore surface.

There are spherical packing particles of uniform size, Figure C.3. We need to evaluate the mass transfer rate both in the interstitial fluid (space between particles) and in the intra-particle void space (pores).

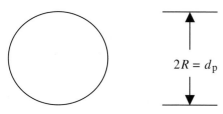

Figure C.3

1 Interstitial fluid:

$$\frac{\partial c}{\partial t} = -\vec{\nabla} \cdot \vec{N} + \frac{\Re}{\alpha},$$

where $\Re$ = rate of mass transfer to particle surface;

$$N_z = c\frac{V_0}{\alpha} - \frac{D_0}{\alpha}\frac{\partial c}{\partial z}$$

$$\alpha\frac{\partial c}{dt} = -V_0\frac{\partial c}{\partial z} + D_0\frac{\partial^2 c}{\partial z^2} + \Re.$$

Now we need to account for the fraction of the packed column which is solid particles:

$$\frac{\text{solid volume}}{\text{total volume}} = \frac{(\text{number of particles}) \times (\text{volume/particle})}{\text{volume}}$$

$$1 - \alpha = \frac{n\left(\frac{4}{3}\pi R^3\right)}{\left(\pi d_p^2/4\right)\Delta z}$$

where n = number of particles in a volume element of column.

$$\Re = -\frac{n4\pi R^2 k_p(c - c_{iR})}{\pi d_p^2 \Delta z/4}$$

where c_{iR} = concentration on particle surface

c_i = concentration in intra-particle void space

k_p = mass transfer coefficient.

Substituting for n:

$$\Re = -\frac{(1-\alpha)(\pi d^2/4\Delta z)}{\frac{4}{3}\pi R^3(\pi d^2/4\Delta z)}4\pi R^2 k_p(c - c_{iR})$$

$$= -\frac{3(1-\alpha)}{R}k_p(c - c_{iR}).$$

Substituting for $\Re$ in differential equation:

$$\alpha\frac{\partial c}{\partial t} = -V_0\frac{\partial c}{\partial z} + D_0\frac{\partial^2 c}{\partial z^2} - \frac{3(1-\alpha)}{R}k_p(c - c_{iR}) \qquad \text{(interstitial fluid equation)}.$$

2 Intra-particle void space (assuming cylindrical pores):

$$k_p(c - c_{iR}) = -D_i \frac{\partial c_i}{\partial r}\bigg|_R \qquad \text{(boundary condition)}$$

$$D_i = \frac{\beta D_{AB}}{q} = \text{effective diffusion coefficient}$$

where β = internal intra-particle void fraction

$\quad q$ = tortuosity

$\quad D_{AB}$ = molecular diffusion coefficient.

$$\beta \frac{\partial c_i}{\partial t} = -\vec{\nabla} \cdot \vec{N}_r + \Re' \qquad \text{intra-particle fluid differential mass balance}$$

$$\vec{N}_r = -D_i \frac{\partial c_i}{\partial r} \qquad \text{neglect convection in pores}$$

$$\Re' = \rho_p \frac{\partial c_{ads}}{\partial t} = \text{mass transfer between fluid and solid phases in particle,}$$

where ρ_p = particle density

$$c_{ads} = \frac{\text{mass adsorbed}}{\text{mass adsorbent (particle)}}$$

$$K_a = \frac{c_{ads}}{c_{i,eq}} = \frac{k_a}{k_d} = \frac{\text{rate of adsorption}}{\text{rate of desorption}} \Rightarrow \text{first-order process}$$

$$\frac{\partial c_{ads}}{\partial t} = k_a(c_i - c_{i,eq}) = k_a\left(c_i - \frac{c_{ads}}{K_a}\right),$$

where K_a is the equilibrium constant and $c_{i,eq}$ is the concentration in the intra-particle void space that would be in equilibrium with c_i.

3 Laplace transform of equation for c_{ads}:

$$s\bar{c}_{ads} = k_a\bar{c}_i - k_a\frac{\bar{c}_{ads}}{K_a}.$$

Using the initial condition: $c_{ads}(r \geq 0, t = 0) = 0$ the solution for $\bar{c}_{ads}$ is:

$$\bar{c}_{ads} = \frac{k_a\bar{c}_i}{s + k_a/K_a}.$$

Now, solving the equations for c and c_i:

Boundary conditions	Initial conditions
$c_i(r = 0) = \text{finite}$	$c(z > 0, t = 0) = 0$
$\dfrac{\partial c_i}{\partial r}\bigg|_{r=0} = 0$	$c_i(r \geq 0, t = 0) = 0$
$c(z = 0, t) = c_0(t)$	
$c(z \to \infty, t) = \text{finite}$	

$$\frac{D_0}{\alpha}\frac{\partial^2 c}{\partial z^2} - \frac{V_0}{\alpha}\frac{\partial c}{\partial z} - \frac{3(1-\alpha)}{R\alpha}k_p(c - c_{iR}) = \frac{\partial c}{\partial t}$$

$$\frac{D_0}{\alpha}\frac{\partial^2 c}{\partial z^2} - \frac{V_0}{\alpha}\frac{\partial c}{\partial z} - \frac{3(1-\alpha)}{R\alpha}\left(D_i\frac{\partial c_i}{\partial r}\bigg|_R\right) = \frac{\partial c}{\partial t} \qquad \text{interstitial fluid equation}$$

$$\beta\frac{\partial c_i}{\partial t} = D_i\frac{1}{r}\frac{\partial}{\partial r}\left(r\frac{\partial c_i}{\partial r}\right) - \rho_p\frac{\partial c_{ads}}{\partial t}$$

$$\frac{D_i}{\beta}\frac{\partial^2 c_i}{\partial r^2} + \frac{2D_i}{\beta r}\frac{\partial c_i}{\partial r} - \frac{\rho_p}{\beta}\frac{\partial c_{ads}}{\partial t} = \frac{\partial c_i}{\partial t} \qquad \text{intra-particle equation.}$$

Take Laplace transform of each equation and substitute:

$$\frac{D_0}{\alpha}\frac{d^2\bar{c}}{dz^2} - \frac{V_0}{\alpha}\frac{d\bar{c}}{dz} - s\bar{c} - \frac{3}{R}D_i\left(\frac{1-\alpha}{\alpha}\right)\frac{d\bar{c}_i}{dr}\bigg|_R = 0$$

$$\frac{D_i}{\beta}\frac{d^2\bar{c}_i}{dr^2} + \frac{2D_i}{\beta r}\frac{d\bar{c}_i}{dr} - s\left(\bar{c}_i + \frac{\rho_p}{\beta}\bar{c}_{ads}\right) = 0$$

$$\frac{D_i}{\beta}\frac{d^2\bar{c}_i}{dr^2} + \frac{2D_i}{\beta r}\frac{d\bar{c}_i}{dr} - s\left(\bar{c}_i + \frac{\rho_p}{\beta}\frac{k_a\bar{c}_i}{s + k_a/K_a}\right) = 0$$

$$r\frac{d^2\bar{c}_i}{dr^2} + 2\frac{d\bar{c}_i}{dr} - b^2 r\bar{c}_i = 0; \qquad b^2 = \frac{s\beta}{D_i}\left(1 + \frac{\rho_p}{\beta}\frac{k_a}{s + k_a/K_a}\right).$$

Boundary conditions become:

$$\bar{c}_i(r = 0, s) = \text{finite}$$

$$\frac{d\bar{c}_i}{dr}\bigg|_{r=0} = 0.$$

To solve the equation for $\bar{c}_i$, we can use a variable substitution.

Let:
$$\bar{c}_i = \frac{u}{r}$$

$$\frac{d\bar{c}_i}{dr} = \frac{u'}{r} - \frac{u}{r^2}$$

$$\frac{d^2\bar{c}_i}{dr^2} = \frac{u''}{r} - \frac{u'}{r^2} - \frac{u'}{r^2} + \frac{2u}{r^3} = \frac{u''}{r} - \frac{2u'}{r^2} + \frac{2u}{r^3}.$$

Substitute:

$$u'' - b^2 u = 0.$$

The solution for u is:

$$u = k_1 e^{br} + k_2 e^{-br}$$

$$\frac{u}{r} = \frac{k_1}{r}e^{br} + \frac{k_2}{r}e^{-br} = \frac{b}{r}(k_1' \sinh br + k_2' \cosh br)$$

(note that we have added a constant (b) and changed k to k').

Applying the boundary conditions:

$$\bar{c}_i(r = 0) \to \text{finite} \Rightarrow k_2' = 0$$

$$\bar{c}_i(r, s) = \frac{k_1'}{r} \sinh br$$

$$\frac{d\bar{c}_i}{dr} = \frac{k_1' b}{r^2} \sinh br + \frac{k_1' b^2}{r} \cosh br = \frac{k_1' b}{r^2}(br \cosh br - \sinh br).$$

Use L'Hopital's rule to check $\lim_{r \to 0}$:

$$\lim_{r \to 0}\left\{ \frac{d\bar{c}_i}{dr} = \frac{k_1' b}{2r}\left(b \cosh br + b^2 r \sinh br - b \cosh br\right)\right\} = 0$$

$$\left. \frac{d\bar{c}_i}{dr}\right|_R = \frac{k_p}{D_i}(\bar{c} - \bar{c}_{iR}) = \frac{k_p}{D_i}\left(\bar{c} - \frac{k_1' b}{R}\sinh br\right).$$

Combine and solve for k_1':

$$k_1' = \frac{R^2 k_p \bar{c}}{D_i b \left[bR \cosh bR + \left(\dfrac{k_p R}{D_i} - 1\right)\sinh bR\right]}.$$

Now we know $\bar{c}_i$ and $d\bar{c}_i/dr$.

Go back to:

$$\frac{D_0}{\alpha}\frac{d^2\bar{c}}{dz^2} - \frac{V_0}{\alpha}\frac{d\bar{c}}{dz} - s\bar{c} - \frac{3}{R}D_i\left(\frac{1-\alpha}{\alpha}\right)\left.\frac{d\bar{c}_i}{dr}\right|_R = 0.$$

Substitute for $\left.\dfrac{d\bar{c}_i}{dr}\right|_R$:

$$\frac{D_0}{\alpha}\frac{d^2\bar{c}}{dz^2} - \frac{V_0}{\alpha}\frac{d\bar{c}}{dz} + A\bar{c} = 0$$

$$A = -\left\{ s + \frac{3}{R}\left(\frac{1-\alpha}{\alpha}\right)k_p\left[1 - \underbrace{\frac{k_p R \sinh bR}{D_i\left[bR \cosh bR + \left(\dfrac{k_p R}{D_i} - 1\right)\sinh bR\right]}}_{h(s)}\right]\right\}.$$

Let $\bar{c}(z, s) = \bar{c}_0(s)e^{\lambda z}$

$$\lambda = \frac{V_0}{2D_0} - \sqrt{\left(\frac{V_0}{2D_0}\right)^2 + \frac{\alpha}{D_0}[s + h(s)]} \quad \text{(note that we only used } -\sqrt{\ } \text{ value).}$$

We can show that:

$$\mu'_1(z) = \mu'_1(0) + \frac{z}{V}(1 + \delta_0)$$

$$\delta_0 = \frac{1-\alpha}{\alpha}\beta\left(1 + \rho_p\frac{K_a}{\beta}\right)$$

$$\mu_2(z) = \mu_2(0) + \frac{2z}{V}\left[\delta_1 + \frac{D_0}{\alpha}\frac{(1+\delta_0)^2}{V^2}\right]; \quad V = \frac{V_0}{\alpha}$$

$$\delta_1 = \underbrace{\left(\frac{1-\alpha}{\alpha}\right)\beta}_{a}\left[\frac{\rho_p}{\beta}\frac{K_a^2}{k_a} + \underbrace{\frac{R^2\beta}{15}\left(1 + \frac{\rho_p}{\beta}K_a\right)^2}_{i}\underbrace{\left(\frac{1}{D_i} + \frac{5}{k_p R}\right)}_{e}\right]$$

δ_a = contribution to concentration profile spreading due to adsorption
δ_i = spreading due to pore diffusion
δ_e = spreading due to mass transfer between particle and interstitial fluid.

Application of the method
1 Perform several pulse injections with variation in V.
2 Plot $\Delta\mu'_1$ vs z/V. The slope (Figure C.4) is $(1 + \delta_0)$. Use the result to obtain a value for K_a.
3 The analysis will depend on the operating range of flowrates. D_0 can depend on V in certain regimes (Figure C.5).

The pulse injection results can be plotted in the form shown in Figure C.6 or, if $D_0 = \gamma V$, as shown in Figure C.7 to determine the operating range.

The result of the above plots is a value for δ_1 which can be obtained from the intercept. We need to perform additional experiments to isolate the contributions of δ_a, δ_i, and δ_e.

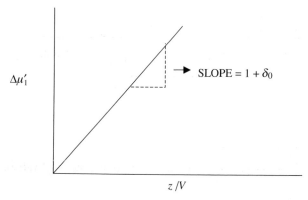

$\Delta\mu'_1$

SLOPE $= 1 + \delta_0$

z/V

Figure C.4

297

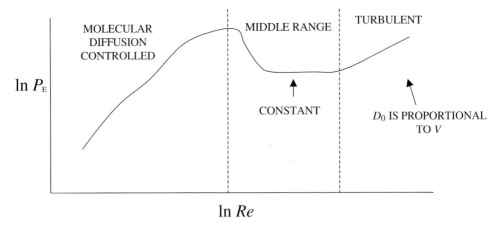

Figure C.5

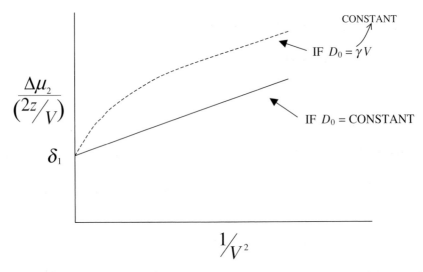

Figure C.6

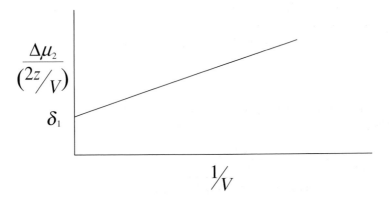

Figure C.7

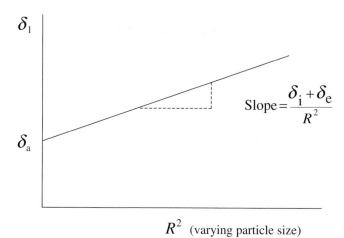

$$\text{Slope} = \frac{\delta_i + \delta_e}{R^2}$$

$$R^2 \text{ (varying particle size)}$$

Figure C.8

We can use a correlation to obtain the parameter k_p:

$$Sh = \frac{d_p k_p}{D_{AB}} = 2.0 + 0.60 Re^{1/2} Sc^{1/3},$$

where

$$D_{AB} = D_i \frac{q}{\beta}.$$

If we now perform experiments where we vary the particle size, we can generate the plot in Figure C.8.

A value for δ_a will allow us to calculate k_a. Likewise, we can obtain D_i from the value of δ_i.

C.2.3 Example C.3: first-order reactions in chromatography

Pulse analysis of chromatography experiments is proposed for a bed packed with solid catalyst particles of uniform diameter. A first-order irreversible reaction occurs at the surface of the spherical particles. How may such a procedure be carried out? (a) Show main steps in the mathematical analysis.

What if the particles are porous? (b) Write an expression for the zeroth moment of an output pulse for this case.

Solution to part (a)
First-order reaction at surface of particles in a packed column.
Reactant balance:

$$\frac{\partial c}{\partial t} = -v \frac{\partial c}{\partial t} + D \frac{\partial^2 c}{\partial z^2} - \underbrace{\frac{3}{R} \frac{(1-\alpha)k_p}{\alpha}(c - c_s)}_{\text{mass transfer to particle surface}},$$

where c_s is the concentration at the surface of the particle.

Surface concentration balance:

$$\frac{\partial c_s}{\partial t} = \frac{3}{R}\frac{(1-\alpha)}{\alpha}k_p(c - c_s) - kc_s \longrightarrow \text{first-order reaction.}$$

Initial conditions:
$$c(t = 0, z) = 0$$
$$c_s(t = 0, z) = 0.$$

Boundary conditions:
$$c(t \leq 0, z = 0) = c_0(t)$$
$$c_s(t \geq 0, z = 0) = 0$$
$$c(t, z \to \infty) = \text{finite.}$$

Laplace transform:

$$\frac{D_0}{\alpha}\frac{\partial^2 \bar{c}}{\partial z^2} - v\frac{\partial \bar{c}}{\partial z} - s\bar{c} - \frac{3}{R}\frac{(1-\alpha)}{\alpha}k_p(\bar{c} - \bar{c}_s) - s\bar{c} + \frac{3}{R}\frac{(1-\alpha)}{\alpha}k_p(\bar{c} - \bar{c}_s) - k\bar{c}_s = 0,$$

yielding:

$$\bar{c}_s = \frac{3}{R}\frac{(1-\alpha)}{\alpha}k_p \Big/ \left[s + k + \frac{3}{R}\frac{(1-\alpha)}{\alpha}k_p\right]$$

$$v = \frac{v_0}{\alpha}.$$

$$\therefore D\bar{c}'' - v\bar{c}' - A\bar{c} = 0,$$

$$\text{where } A = \frac{3}{R}\frac{(1-\alpha)}{\alpha}k_p\left[1 - \frac{\left(\dfrac{3}{R}\dfrac{(1-\alpha)}{\alpha}k_p\right)}{\left(s + k + \dfrac{3}{R}\dfrac{(1-\alpha)}{\alpha}k_p\right)}\right] + s \qquad \nearrow \text{ from } \frac{\partial c}{\partial t} \text{ term.}$$

Boundary conditions: $\bar{c}(z = 0) = \bar{c}_0(s)$
$$\bar{c}(z \to \infty) = \text{finite.}$$

Solution to part (b)

$$\bar{c}(z) = \bar{c}_0 e^{\lambda z} \quad \text{where } \lambda = \frac{1}{2D_0}\left[v_0 \pm \sqrt{v_0^2 + 4D_0 A}\right]$$

$$m_0(z) = \lim_{s \to 0}\bar{c}(z) = \underbrace{(\lim_{s \to 0}\bar{c}_0)(\lim_{s \to 0}e^{\lambda z})}_{m_0(z = 0)}$$

$$\lim_{s \to 0}\lambda = \frac{1}{2D_0}\left[v_0 - \sqrt{v_0^2 + 4D_0 \lim_{s \to 0}A}\right] \equiv \lambda_0 \leq 0$$

$$\lim_{s \to 0}A = \frac{3}{R}\frac{(1-\alpha)}{\alpha}k_p\left[1 - \frac{k_p\dfrac{3}{R}\dfrac{(1-\alpha)}{\alpha}}{k + \dfrac{3}{R}\dfrac{(1-\alpha)}{\alpha}k_p}\right].$$

$$\therefore m_0(z) = m_0(z = 0)e^{\lambda z}.$$

∴ Zeroth moment decreases with z since solute is lost in chemical reaction.

$$m_1(z) = -\lim_{s\to 0} \frac{d\bar{c}}{ds},$$

where $\dfrac{d\bar{c}}{ds} = \dfrac{d\bar{c}_0}{ds}e^{\lambda z} + \bar{c}_0 z e^{\lambda z}\dfrac{d\lambda}{ds}$

and $\dfrac{d\lambda}{ds} = \dfrac{-1/2}{2D_0}(v_0^2 + D_0 A)^{-1/2} 4D_0 \dfrac{dA}{ds}.$

Hence,

$$\left[\begin{array}{l} -m_1(z) = -m_1(0)e^{\lambda_0 z} + m_0(0)z e^{\lambda_0 z}\gamma \\[2mm] \mu_1^*(z) = \mu_1(0)e^{\lambda_0 z} - z\gamma^{\lambda_0 z} \end{array}\right] \quad \text{solution in terms of moments.}$$

$$\frac{dA}{ds} = \left\{ \frac{-\left[\dfrac{3}{R}\dfrac{(1-\alpha)}{\alpha}k_p\right]^2}{\left[s + k + \dfrac{3}{R}\dfrac{(1-\alpha)}{\alpha}k_p\right]^2} \right\} + 1$$

$$\lim_{s\to 0}\frac{d\lambda}{ds} = \gamma < 0.$$

Exit of pulse concentration gradient (C.G.) is retarded by loss of reactant. Note that if we are to get an output pulse, k must be very small so that we don't lose all reactant, or z must be small.

From before: $K_a \to \infty \Rightarrow k_a = 0$ or non-reversible first-order reaction but $k_a \neq 0$

$$\frac{\partial c_i}{\partial t} = \frac{D_i}{\beta}\left(\frac{\partial^2 c_i}{\partial r^2} + \frac{2}{r}\frac{\partial c_i}{\partial r}\right) - \frac{\rho_p}{\beta}k_a c_i$$

$$\frac{D_i}{\beta}\bar{c}_i'' + \frac{2D_i}{r\beta}\bar{c}_i' - \bar{c}_i\left(\frac{\rho_p}{\beta}k_a + s\right) = 0$$

$$r\bar{c}_i'' + 2\bar{c}_i'' - \underbrace{\frac{\beta}{D_i}\left[s + \frac{\rho_p}{\beta}k_a\right]}_{\equiv\, b^2} r\bar{c}_i = 0.$$

Note: $\displaystyle\lim_{s\to 0} = \sqrt{\frac{k_a\rho_p}{D_i}} \equiv b_0.$

$$\bar{c}(z) = \bar{c}_0 e^{\lambda z} \quad \text{where} \quad \lambda = \frac{v_0}{2D_0} - \sqrt{\left(\frac{v_0}{2D_0}\right)^2 + \frac{\alpha}{D_0}[s + h(s)]}$$

and

$$h(s) = \frac{3k_p}{R} \frac{(1-\alpha)}{\alpha} \left\{ \frac{-\sinh bR + bR \cosh bR}{bR \cosh bR + \left(1 - \dfrac{k_p R}{D_i}\right) \sinh bR} \right\}$$

$$m_0(z) = \lim_{s \to 0} \bar{c}(z) = \underbrace{(\lim_{s \to 0} \bar{c}_0)(\lim_{s \to 0} e^{\lambda z})}_{m_0(0)}$$

$$\lim_{b \to b_0} h(s) = \frac{3k_p}{R} \frac{(1-\alpha)}{\alpha} \left\{ \frac{-\sinh b_0 R + b_0 R \cosh b_0 R}{bR \cosh bR + \left(1 - \dfrac{k_p R}{D_i}\right) \sinh Rb_0} \right\} \equiv h_0$$

$$h_0(z) = m_0(0)e^{\lambda_0 z} \text{ where } \lambda_0 \equiv \frac{v_0}{2D_0} - \sqrt{\left(\frac{v_0}{2D_0}\right)^2 + \frac{\alpha}{D_0} h_0} \le 0.$$

$\therefore m_0(z)$ decreases with increasing z.

C.3   Question

C.1 Breakthrough curves in ion-exchange columns. Experiments are proposed for a column packed with H^+/Na^+ ion-exchange resin. The resin has been totally regenerated with acid. A step function of sodium-chloride solution is injected into the column at time $t = 0$. Sketch output curves of concentration vs time for the Cl^-, Na^+, and H^+ ions. It is proposed that only a pH meter at the outlet is needed to get information about the breakthrough curves and of pore diffusion and kinetics. Discuss this possibility and its significance.

Appendix D: Finite difference approach

Equilibrium-stage processes are discrete steps. One approach to the analysis is an evaluation as a finite difference calculation where each stage is an equal and discrete interval in the process train. Obviously, a process simulator can be used. Finite difference approaches, including ones shown here, can be implemented on spreadsheets for rapid estimates.

Initially, we can define some finite difference mathematical operations. We will then demonstrate how the approach is applied to equilibrium-stage separation processes.

D.I   Definitions

Given a function $y = f(x)$, we can define the value at the nth interval point as:

$$y_n = f(x_0 + n\Delta x) \qquad \text{where } \Delta x = \text{discrete interval which is a constant}$$
$$x_0 = \text{initial value of } x.$$

The difference in the value of the function y between two interval points can be described as a first forward difference:

$$\Delta y_n = f(x_0 + (n+1)\Delta x) - f(x_0 + n\Delta x) = y_{n+1} - y_n.$$

We can continue this operation:

$$\Delta(\Delta y_n) = \Delta^2 y_n = \text{second forward difference}$$
$$= \Delta(y_{n+1} - y_n) = (y_{n+2} - y_{n+1}) - (y_{n+1} - y_n)$$
$$= y_{n+2} - 2y_{n+1} + y_n.$$

This can be generalized to write an expression for the kth forward difference:

$$\Delta^k y_n = \sum_{r=0}^{k} (-1)^r \frac{k!}{r!(k-r)!} y_{n+k-r}.$$

We can next define a shifting operator (E) which moves the index forward:

$$E = 1 + \Delta$$
$$E y_n = (1 + \Delta) y_n = y_n + y_{n+1} - y_n = y_{n+1}.$$

Again, generalizing

$$E^k y_n = y_{n+k}.$$

There are three rules which can also be useful.

Index rule

$$\Delta^r \Delta^s y_n = \Delta^s \Delta^r y_n = \Delta^{r+s} y_n.$$

Commutative rule

$$\Delta(C y_n) = C y_{n+1} - C y_n = C \Delta y_n \qquad C = \text{constant}.$$

Distributive rule

$$\Delta(y_n + z_n) = \Delta y_n + \Delta z_n.$$

Equilibrium-stage mass balances give rise to general linear difference equations. For a difference equation of order k:

$$y_{n+k} + P_{n+k-1} y_{n+k-1} + \cdots + P_{n+1} y_{n+1} + P_n y_n = Q,$$

where each P, and Q, are constants.

The complete solution includes a homogeneous and particular component:

$$y_n = y_n^h + y_n^P.$$

Using the shifting operator, we can rewrite the homogeneous difference equation as:

$$(E^k + P_{n+k-1} E^{k-1} + \cdots + P_{n+1} E^1 + P_n E^0) y_n = 0.$$

This equation can be factored for the k roots:

$$(E - \alpha_1)(E - \alpha_2) \ldots (E - \alpha_k) y_n = 0.$$

Each root can be evaluated:

$$(E - \alpha_i) y_n = 0 \Rightarrow y_n + 1 - \alpha_i y_n = 0.$$

Assume a solution of the form: $y_n = C\beta^n$.
Substituting:

$$C\beta^{n+1} - \alpha_i C\beta^n = 0 \Rightarrow \beta = \alpha_i;$$
$$\therefore y_n = C\alpha_i^n.$$

The total homogeneous solution becomes:

$$y_n^h = \sum_{i=1}^{k} C_i \alpha_i^n.$$

If two roots are equal (α_j, for example):

$$y_n = \alpha_j(C_1 n + C_2).$$

D.2 Application to separation process

Now, let us apply the mathematical approach to an equilibrium-stage process illustrated schematically in Figure D.1.

A mass balance around stage $(n + 1)$, assuming L and V are constant, gives:

$$V y_{n+2} + L x_n = V y_{n+1} + L x_{n+1}.$$

We further assume: $y_{n+1} = m x_{n+1}$ (linear equilibrium relation).
The mass balance equation becomes:

$$y_{n+2} - (1 + A)y_{n+1} + A y_n = 0; \quad A = \frac{L}{mV}.$$

Solution is of form: $y_n = C\alpha^n$.
Substituting:

$$C\alpha^{n+2} - (1 + A)C\alpha^{n+1} + AC\alpha^n = 0.$$

Dividing by $C\alpha^n$:

$$\alpha^2 - (1 + A)\alpha + A = 0.$$

The two values of which satisfy this equation are $\alpha = 1, A$.
Therefore: $y_n = C_1 A^n + C_2$.
The constants C_1 and C_2 are evaluated from boundary conditions on the equilibrium-stage process. They remain constant for each stage.

D.2.1 Example D.1

Problem:
Derive the Kremser equation, Equation (3.49). Refer to Figure 3.27 for a diagram of the equilibrium-stage process.

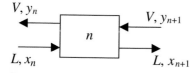

Figure D.1 Schematic of equilibrium-stage process.

Solution:

The boundary conditions are:

$$n = 0; \qquad x = x_0$$
$$n = N + 1; \quad y = y_{N+1}.$$

The general solution is $y_n = C_1 A^{n+1} + C_2$.

In Chapter 2, we defined:

$$y_1^* = mx_0 \text{ (value of } y \text{ which would be in equilibrium with } x_0)$$
$$= y_0$$
$$y_0 = y_1^* = C_1 + C_2$$
$$y_{N+1} = C_1 A^{n+1} + C_2.$$

We can immediately write:

$$\frac{y_{N+1} - y_1}{y_{N+1} - y_1^*} = \frac{C_1(A^{n+1} - A)}{C_1(A^{n+1} - 1)} = \frac{A - A^{n+1}}{1 - A^{n+1}}.$$

This is a similar equation to that given in Section 3.7.

D.2.2 Example D.2: distillation, separation of $H_2{}^{18}O$ from $H_2{}^{16}O$

Problem:

Water contains 0.002 mole fraction $H_2{}^{18}O$. Using the Fenske equation, estimate the number of stages required to separate $H_2{}^{18}O$ from $H_2{}^{16}O$ and compare to the result from the Smoker equation.

Overall mass balance on entire column section (Figure D.2):

$$V y_{n-1} = L x_n + D x_D.$$

Relative volatility: $\alpha = \dfrac{K_A}{K_B} = \dfrac{y(1 - x)}{x(1 - y)}$, assumed constant.

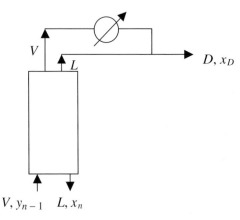

V

L

D, x_D

V, y_{n-1} L, x_n

Figure D.2 Enriching section of distillation column.

Rearranging,

$$y = \frac{\alpha x}{1 + x(\alpha - 1)} \quad \text{applies to any stage.}$$

Substituting into the overall mass balance:

$$\frac{\alpha x_{n-1}}{1 + x_{n-1}(\alpha - 1)} = \frac{L}{V} x_n + \frac{D}{V} x_D; \quad D = V - L;$$

$$\frac{D}{V} = 1 - \frac{L}{V} = 1 - R_V = R_V x_n + (1 - R_V) x_D; \quad R_V = \frac{L}{V} = \text{internal reflux ratio.}$$

$$x_n x_{n-1} + a x_n + b x_{n-1} + C = 0 \text{ (Ricatti difference equation);}$$

$$a \equiv \frac{1}{\alpha - 1}; \quad b \equiv \frac{(\alpha - 1) x_D (1 - R_V) - \alpha}{R_V(\alpha - 1)}; \quad c \equiv \frac{x_D(1 - R_V)}{R_V(\alpha - 1)}.$$

Substitution: let $x_n = z_n + h$ (this is a coordinate transformation which shifts the system to the intersection of the operating and equilibrium lines):

$$(z_n + h)(z_{n-1} + h) + a(z_n + h) + b(z_{n-1} + h) + c = 0$$

$$z_n z_{n-1} + z_n(a + h) + z_{n-1}(b + h) + \underbrace{h^2 + h(a + b) + c}_{\text{set} = 0} = 0$$

$$h = \frac{1}{2}\left[-(a + b) \pm \sqrt{(a + b)^2 - 4c} \right] \Rightarrow \text{becomes defining equation for } h.$$

Now divide by $z_n z_{n-1}$:

$$1 + \frac{1}{z_{n-1}}(a + h) + \frac{1}{z_n}(b + h) = 0.$$

Let $v_n = \frac{1}{z_n}$

$$v_n(b + h) + v_{n-1}(a + h) = -1$$

$$v_n^{(h)} = c\beta^n$$

$$\beta^n(b + h) + \beta^{n-1}(a + h) = 0$$

$$\beta = -\frac{a + h}{b + h}$$

$$v_n^{(P)} = -\frac{1}{a + b + 2h}$$

$$v_n = C_1\left(-\frac{a + h}{b + h} \right)^n - \frac{1}{a + b + 2h} = \frac{1}{z_n} = \frac{1}{x_n - h}.$$

Rearrange and substitute to get the equation for x_n:

$$x_n = h + \frac{1}{C_1\left(-\dfrac{a + h}{b + h} \right)^n - \dfrac{1}{a + b + 2h}}.$$

Boundary condition: $x_m = x_F$ (feed composition) for $n = 0$ (note: this only counts stages in enriching section):

$$C_1 = \frac{1}{x_F - h} + \frac{1}{a + b + 2h}.$$

Smoker equation: $x_n = h + \dfrac{a + b + 2h}{\left(\dfrac{a + b + h + x_F}{x_F - h}\right)\left(-\dfrac{a + h}{b + h}\right)^n - 1}$ in enriching section.

[This equation can also be modified for a stripping section: x_n is replaced by x_B, R_V by the reboiler ratio.]

The stage number is from feed plate down instead of feed plate up.
⎵ stripping section   ⎵ enriching section

Let x_e, y_e be the interaction of the operating and equilibrium lines:

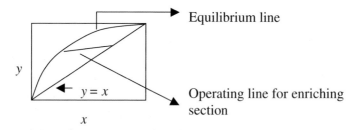

Equilibrium line

y

$y = x$

Operating line for enriching section

x

Equilibrium curve: $y = \dfrac{\alpha x}{1 + x(\alpha - 1)}$

Operating line: $y = x R_V + x_D (1 - R_V)$

Equating: $x_e^2 + x_e(a + b) + c = 0$ (a, b, c defined previously)

$x_e = h$ (see defining equation for h).

$\therefore h$ must be positive, b negative and greater than a or c.

The variable h shifts the coordinate system to the intersection of the operating and stripping curves. One special case is that of total reflux. In this case $D = 0$ as no product is being withdrawn and the reflux ratio $R_V =$ becomes 1. This also corresponds to the minimum number of equilibrium stages required for a separation when this is the case,

$$a = \frac{1}{\alpha - 1}; \quad b = \frac{-\alpha}{\alpha - 1}; \quad c = 0$$

$$x_n = \frac{1}{C_1\left(\dfrac{1}{\alpha}\right)^n + 1}.$$

When $n = 0$, $x_n = x_B$ (composition at bottom of enriching section):

$$C_1 = \frac{1}{x_B} - 1.$$

When $n = N$, $x_n = x_D$:

$$N = \frac{\ln \dfrac{x_D(1 - x_B)}{x_B(1 - x_D)}}{\ln \alpha} \qquad \text{Fenske equation;}$$

$\alpha = 1.006 \qquad$ for H_2O^{16} and H_2O^{18};

and water contains 0.002 mole fraction H_2O^{18}. Therefore:

$$x_D = 0.998$$
$$x_B = 0.9746$$
$$N = 430 \text{ stages} \qquad \text{(from Fenske equation)}$$
$$x_F/x_B = 3920; \qquad N = 503 \text{ stages} \quad \text{(from Smoker equation).}$$

D.3 Problems

D.1 Derive the Kremser equation for the case where $E_{mV} \neq 1$.

D.2 Steam distillation is used to remove ethanol from water. The water enters with an ethanol mole fraction equal to 0.02 and the exit requirement is a mole fraction equal to 10^{-4}. The equilibrium relation is $y = 9x$. For an L/V ratio equal to 5, what is the exit ethanol vapor mole fraction? How many equilibrium stages are needed?

D.3 The flowsheet below is a liquid–liquid extraction process for uranyl nitrate (UN) extraction using tributyl phosphate (TBP). For this problem, the flowrates and equilibrium relation are in lb_m instead of moles. The quantities x and y are the mass

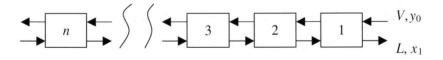

fraction of UN in the L (H_2O) and V (TBP) phase respectively. For conditions where $L = 90 \ lb_m/hr$ and $V = 150 \ lb_m/hr$, derive an equation for y_n in terms of n, given: $x_1 = 1.2 \times 10^{-3}$ and $y_n = 5.5x_n$.

D.4 Redo Example 5.1 cross-flow cascade, using the finite difference approach.

Appendix E: Bibliography of chemical separations and related physical properties

E.1 Books on separations in general

Berg, E. W., *Physical and Chemical Methods of Separation* (New York: McGraw-Hill, 1963).

Henley, E. J. and J. D. Seader, *Equilibrium Stage Operations in Chemical Engineering* (New York: John Wiley and Sons, 1981).

Humphrey, J. L. and G. E. Keller II, *Separation Process Technology* (New York: McGraw-Hill, 1997).

Karger, B. L., L. R. Snyder, and C. Horvath, *An Introduction to Separation Science*. (New York: John Wiley and Sons, 1973).

King, C. J., *Separation Processes*, 2nd edn (New York: McGraw-Hill, 1980).

Li, N. N., ed., *Recent Developments in Separation Science*, multiple volumes (New York: CRC Press, 1972 and subsequent).

Miller, J. M., *Separation Methods in Chemical Analysis* (New York: John Wiley and Sons, 1975).

Minczewski, J. *et al.*, *Separation and Preconcentration Methods in Inorganic Trace Analysis* (New York: John Wiley and Sons, 1982).

Rousseau, R. W., *Handbook of Separation Process Technology* (New York: Wiley–Interscience, 1986).

Schweitzer, P. A., *Handbook of Separation Techniques for Chemical Engineers* (New York: McGraw-Hill, 1979).

Wankat, P. C., *Equilibrium Staged Separations* (New Jersey: Prentice-Hall, 1988).
  Rate-Controlled Separations (Kluwer Academic Publishers, 1990).

Watson, J. S., *Separation Methods for Waste and Environmental Applications* (Marcel-Dekker Pub. Co., 1999).

Weissberger, A. and E. S. Perry, eds., *Techniques of Chemistry: Separation and Purification. Techniques of Chemistry Series*, Vol. 12 (New York: John Wiley and Sons, 1978).

Wolf, F. J., *Separation Methods in Organic Chemistry and Biochemistry* (New York: Academic Press, 1969).

Young, R. S., *Separation Procedures in Inorganic Analysis* (Bucks, England: Charles Griffing and Company, Ltd., 1980; New York: Wiley–Halsted, 1980).

E.2 Books on specific separation techniques

Astarita, G., D. W. Savage, and A. Bisio, *Gas Treating with Chemical Solvents* (New York: Wiley-Interscience, 1983).

Bhave, R. R., *Inorganic Membranes* (New York: Chapman and Hall, 1991).

Giddings, J. C., E. Grushka, J. Cazes, and P. R. Brown, eds., *Advances in Chromatography*, multiple volumes (New York: Marcel-Dekker, 1965 and subsequent).

Hanson, C., T. C. Lo, and M. H. I. Baird, eds., *Solvent Extraction Handbook* (New York: John Wiley and Sons, 1983).

Helfferich, F., *Ion Exchange* (New York: McGraw-Hill, 1962).

Ho, W. S. W. and K. K. Sirkar, *Membrane Handbook* (New York: Chapman and Hall, 1992).

Holland, C. D., *Fundamentals of Multicomponent Distillation* (New York: McGraw-Hill, 1981).

Kohl, A. and F. C. Riesenfeld, *Gas Purification*, 4th edn (Houston: Gulf Publishing, 1985).

Lemlich, R., *Adsorptive Bubble Separation Techniques* (Orlando: Academic Press, 1972).

Marinsky, J. A. and Y. Marcus, eds., *Ion Exchange and Solvent Extraction*, multiple volumes (New York: Marcel-Dekker, 1966 and subsequent).

Mujumdar, A. S., ed., *Advances in Drying*, multiple volumes (New York: Hemisphere Publishing, 1980 and subsequent).

Mulder, M., *Basic Principles of Membrane Technology* (Boston: Kluwer Academic, 1991).

Noble, R. D. and S. A. Stern, *Membrane Separations Technology: Principles and Applications* (New York: Elsevier, 1995).

Ritcey, G. M. and A. W. Ashbrook, *Solvent Extraction: Principles and Applications to Process Metallurgy, Parts I and II* (New York: Elsevier, 1983, 1984).

Ruthven, D. M., *Principles of Adsorption and Adsorption Processes* (New York: John Wiley and Sons, 1984).

Yang, R. T., *Gas Separations by Adsorption Processes* (Boston: Butterworth, 1987).

E.3 Additional bibliography

Clark, M. M., *Transport Modeling for Environmental Engineers and Scientists* (New York: Wiley, 1996).

Cussler, E. L., *Diffusion: Mass Transfer in Fluid Systems* (Cambridge University Press, 1984).

Geankoplis, C. J., *Transport Processes and Unit Operations*, 3rd edn (Boston: Prentice-Hall, 1993).

Kirk-Othmer Encyclopedia of Chemical Technology, 3rd edn (New York: Wiley–Interscience, 1978–1984).

Middleman, S., *An Introduction to Mass and Heat Transfer* (New York: John Wiley and Sons, 1998).

Perry, R. H. and D. Green, eds., *Perry's Chemical Engineers' Handbook*, 6th edn (New York: McGraw-Hill, 1984).

Reynolds, T. D., *Unit Operations and Processes in Environmental Engineering* (Boston: PWS Publishing Co., 1982).

Sherwood, T. K., R. L. Pigford, and C. R. Wilke, *Mass Transfer* (New York: McGraw-Hill, 1975).

Weber Jr., W. J. and F. A. DiGiano, *Process Dynamics in Environmental Systems* (New York: Wiley, 1996).

E.4   Phase equilibrium

There are a number of sources for phase equilibrium data and computational methods (see E4.1, below). Most of the material focuses on vapor–liquid equilibrium (VLE) since this information is used extensively for distillation, absorption, and stripping. The most complete VLE literature is a series of books by Hala *et al.* (1967, 1968). Additional information can be found in Hirata *et al.* (1975) and Gmehling *et al.* (1979). For light hydrocarbon systems, the Natural Gas Processors Association has published a data book (1972). A very useful and extensive source, including solid–liquid and liquid–liquid as well as VLE information, has been written by Walas (1985). This book contains both source data and methodology and contains sample calculations.

When no data are available, there are simulation packages available that can provide estimates. ASPEN and HYSYS are two popular ones. The thermodynamics book by Sandler (1989) contains a diskette that can be used for phase equilibrium calculations. Additional texts are Prausnitz *et al.* (1980), Prausnitz *et al.* (1986), and Reid *et al.* (1987).

There are some published articles that contain data and calculation procedures: Yaws *et al.* (1990, 1993, 1995).

E.4.1   Bibliography

Chu, J. C., R. J. Getty, L. F. Brennecke, and R. Paul, *Distillation Equilibrium Data* (New York: Reinhold, 1950).

Engineering Data Book (Tulsa, OK: Natural Gasoline Supply Men's Association, 421 Kennedy Bldg, 1953).

Gmehling, J., U. Onken, and W. Arlt, *Vapor–Liquid Equilibrium Collection* (continuing series, Frankfurt: DECHEMA, 1979–).

Hala, E., J. Pick, V. Fried, and O. Vilim, *Vapor–Liquid Equilibrium*, 2nd edn (Oxford: Pergamon, 1967).

Hala, E., I. Wichterle, J. Polak, and T. Boublik, *Vapor–Liquid Equilibrium at Normal Pressures* (Oxford: Pergamon, 1968).

Hirata, M., S. Ohe, and K. Nagahama, *Computer Aided Data Book of Vapor–Liquid Equilibria* (Amsterdam: Elsevier, 1975).

Horsely, L. H., *Azeotropic Data, ACS Advances in Chemistry*, No. 6 (Washington, DC: American Chemical Society, 1952).

Azeotropic Data (II), ACS Advances in Chemistry, No. 35 (Washington, DC: American Chemical Society, 1952).

Maxwell, J. B., *Data Book on Hydrocarbons* (Princeton, NJ: Van Nostrand, 1950); *Engineering Data Book*, 9th edn (Tulsa, OK: Natural Gas Processors Suppliers Assn., 1972).

Perry, R. H. and D. Green, eds., *Perry's Chemical Engineer's Handbook*, 6th edn (New York: McGraw-Hill, 1984).

Prausnitz, J. M., T. F. Anderson, E. A. Grens, C. A. Eckert, R. Hsieh, and P. O'Connell, *Computer Calculations for Multicomponent Vapor–Liquid and Liquid–Liquid Equilibria* (Englewood Cliffs, NJ: Prentice-Hall, 1980).

Prausnitz, J. M., R. N. Lichtenthaler, and E. G. de Azevedo, *Molecular Thermodynamics of Fluid-Phase Equilibria*, 2nd edn (Englewood Cliffs, NJ: Prentice-Hall, 1986).

Reid, R. C., J. M. Prausnitz, and B. E. Poling, *The Properties of Gases and Liquids*, 4th edn (New York: McGraw-Hill, 1987).

Sandler, S. I., *Chemical and Engineering Thermodynamics*, 2nd edn (New York: Wiley, 1989).

Timmermans, J., *The Physico-Chemical Constants of Binary Systems in Concentrated Solutions*, five vols. (New York: Interscience, 1959–60).

Walas, S. M., *Phase Equilibria in Chemical Engineering* (Reading, MA: Butterworths, 1985).

Wichterle, I., J. Linek, and E. Hala, *Vapor–Liquid Equilibrium Data Bibliography* (Amsterdam: Elsevier, 1973).

Yaws, C. L., H.-C. Yang, J. R. Hopper, and K. C. Hansen, 232 hydrocarbons: water solubility data, *Chem. Engg*, April 1990, 177–181.

Yaws, C. L., X. Pan, and X. Liu, Water solubility data for 151 hydrocarbons, *Chem. Engg*, Feb. 1993, 108–111.

Yaws, C. L., L. Bu, and S. Nijhawan, Calculate the solubility of aromatics, *Chem. Engg*, Feb. 1995, 113–115.

References

CHAPTER 1

1 National Research Council, *Separation & Purification: Critical Needs and Opportunities* (National Academy Press, 1987).
2 *Chemical and Engineering News*, December 1, 1997.
3 Byers, Charles H. and Ammi Amarnath, Understand the potential of electro-separations, *Chem. Eng. Prog.*, February 1995.
4 Zofnass, P., *1993 National Congress for the Advancement of Minorities in the Environmental Professions*, Washington, DC, February 24–26, 1993.
5 Holmes, G., B. R. Singh, and L. Theodore, *Handbook of Environmental Management and Technology* (J. Wiley & Sons, 1993).
6 Chemical Manufacturers Association, *Designing Pollution Prevention into the Process*, © CMA (1993).
7 Mulholland, K. L. and J. A. Dyer, Reduce waste and make money, *Chem. Eng. Prog.*, January 2000.
8 Environment Canada website on acid rain – http://www.ns.ec.gc.ca/aeb/ssd/acid/acidfaq.html
9 Cowling, E. B., Acid rain in historical perspective, *Environ. Sci. Tech.*, **16**(2) (1982).

CHAPTER 2

1 Middleman, S., *An Introduction to Mass and Heat Transfer* (New York: John Wiley and Sons, 1998).
2 Keller, G., *Separations: New Directions for an Old Field*, AIChE Monograph Series, Vol. 83, No. 17 (New York: AIChE, 1987).
3 Null, H. R., *Handbook of Separation Processes*, Chap. 22, R. W. Rousseau, ed. (New York: Wiley–Interscience, 1986).
4 Wankat, P. C., *Equilibrium Staged Separations* (New Jersey: Prentice-Hall, 1988).
5 Wankat, P. C., *Rate-Controlled Separations* (London: Elsevier, 1990).

6 Belhatecke, D. H., Choose appropriate wastewater treatment technologies, *Chem. Eng. Prog.*, August 1995, pp. 32–51.

7 Zinkus, C. A., W. D. Byers, and W. W. Doerr, Identify appropriate water reclamation technologies, *Chem. Eng. Prog.*, May 1998, pp. 19–31.

8 Fitch, B., Choosing a separation technique, *Chem. Eng. Prog.*, December 1974, pp. 33–37.

9 King, C. J., *Handbook of Separation Processes*, Chap. 15, R. W. Rousseau, ed. (New York: Wiley–Interscience, 1986).

10 Hyman, M. and L. Bagaasen, Select a site cleanup technology, *Chem. Eng. Prog.*, August 1997, pp. 22–43.

11 King, C. J., *Separation Processes*, 2nd edn (New York: McGraw-Hill, 1980).

12 King, C. J., From unit operations to separation processes, *Sep. Pur. Methods*, **29**(2), 2000, pp. 233–245.

CHAPTER 3

1 Wilson equation: G. M. Wilson, *J. Amer. Chem. Soc.*, **86**, 127 (1964).

2 NRTL equation: H. Renon and J. M. Prausnitz, *AIChE J*, **14**, 135 (1968).

3 UNIQUAC equation: D. S. Abrams and J. M. Prausnitz, *AIChE J*, **21**, 116 (1975).

4 David Clough, personal communication.

5 Yang, R. T., *Gas Separation by Adsorption Processes* (Boston: Butterworths, 1987).

6 Tien, C., *Adsorption Calculations and Modeling* (Boston: Butterworth–Heinemann, 1994).

7 Ruthven, D. M., *Principles of Adsorption and Adsorption Processes* (New York: J. Wiley and Sons, 1984).

8 Dobbs, R. A. and J. M. Cohen, *Carbon Adsorption Isotherms for Toxic Organics*, U.S. Environmental Protection Agency, EPA 600/8-80-023.

9 Knaebel, K. S., For your next separation consider adsorption, *Chem. Engg*, **102**(11), November 1995, pp. 92–102.

10 Seader, J. D. and E. J. Henley, *Separation Process Principles* (J. Wiley and Sons, Inc., 1998).

11 King, C. J., *Separation Processes* (McGraw-Hill, 1971).

12 Cussler, E. L., *Diffusion: Mass Transfer in Fluid Systems* (Cambridge University Press, 1984).

13 Bird, R. B., W. E. Stewart, and E. N. Lightfoot, *Transport Phenomena*, 2nd edn (J. Wiley and Sons, Inc., 2002).

14 Middleman, S., *An Introduction to Mass and Heat Transfer* (New York: John Wiley and Sons, 1998).

CHAPTER 4

1 Humphrey, J. L., Separation processes playing a critical role, *Chem. Eng. Prog.*, **91**(10), October 1995, pp. 31–42.

2 Wankat, P. C., *Equilibrium Staged Separations* (New Jersey: Prentice-Hall, 1988).

3 Geankpolis, C. J., *Transport Processes and Unit Operations*, 3rd edn (Boston: Prentice-Hall, 1993).

4 Los Alamos National Laboratories, written communications.

CHAPTER 5

1 McCabe, W. L., J. C. Smith, and P. Harriott, *Unit Operations of Chemical Engineering* (New York: McGraw-Hill Book Company, 1985).
2 *Transactions AIChE*, **36**, 1940, pp. 628–629.
3 Wankat, P. C., *Equilibrium Staged Separations* (New Jersey: Prentice-Hall, 1988).
4 Foust, A. S., L. A. Wenzel, C. W. Clump, L. Maus, and L. B. Anderson, *Principles of Unit Operations*, 2nd edn (New York: Wiley and Sons, 1980).

CHAPTER 6

1 *Chem. Eng. Prog.*, April 1996, p. 56.
2 Kohl, A. L., Absorption and stripping, *Handbook of Separation Process Technology*, R. W. Rousseau, ed. (New York: John Wiley and Sons, 1987), pp. 340–404.
3 Lee, S.-Y. and Y. P. Tsui, Succeed at gas–liquid contacting, *Chem. Eng. Prog.*, July 1999, pp. 23–49.
4 Perry, R. H., C. H. Chilton, and S. O. Kirkpatrick, eds. *Chemical Engineering Handbook*, 4th edn (McGraw-Hill, 1963).
5 Foust, A. S., L. A. Wenzel, C. W. Clump, L. Maus, and L. B. Anderson, *Principles of Unit Operations*, 2nd edn (New York: John Wiley and Sons, 1980).
6 Wankat, P. C., *Equilibrium Stage Separations* (New Jersey: Prentice-Hall, 1988).
7 Seader, J. D. and E. J. Henley, *Separation Process Principles* (John Wiley and Sons, Inc., 1998).
8 Shaeiwitz, J. A., written communications.

CHAPTER 7

1 Keller II, G. E., Adsorption: building upon a solid foundation, *Chem. Eng. Prog.*, **91**(10), October 1995, pp. 56–67.
2 Randey, R. A. and S. Malhotra, Desulfurization of gaseous fuels with recovery of elemental sulfur: an overview, *Critical Reviews in Environmental Science & Technology*, **29**(3), 1999, pp. 237–238.
3 Loretta Y. and F. Li, "Heavy metal sorption and hydraulic conductivity studies using three types of bentonite admixes, *J. Environ. Engg*, **127**(5), May 2001, pp. 420–423.
4 Yang, R. T., *Gas Separation by Adsorption Processes* (Boston: Butterworths, 1987).
5 Ruthven, D. M., Zeolites as selective adsorbents, *Chem. Eng. Prog.*, **84**(2), February 1988, pp. 42–50.
6 Knaebel, K. S., For your next separation consider adsorption, *Chem. Eng. Prog.*, **102**(11), November 1995, pp. 92–102.
7 Keller II, G. E., R. A. Anderson, and C. M. Yon, Adsorption, *Handbook of Separation Process Technology*, R. W. Rousseau, ed. (New York: Wiley, 1987), pp. 644–696.
8 Keller II, G. E., *Separations, New Directions for an Old Field*, AIChE Monograph Series, Vol. 83, No. 17 (New York: AIChE, 1987).
9 Marcus, B. K. and W. E. Cormier, Going green with zeolites, *Chem. Eng. Prog.*, **95**(6), June 1999, pp. 47–53.

10 Basmadjian, D., *The Little Adsorption Book: A Practical Guide for Engineers and Scientists* (CRC Press, 1996).

11 Thomas, H. C., Chromatography: a problem in kinetics, *Annals New York Acad. Sci.*, **49** (1948), pp. 161–182.

CHAPTER 8

1 Streat, M. and F. L. D. Cloete, Ion exchange, *Handbook of Separation Process Technology*, R. W. Rousseau, ed. (New York: Wiley, 1987), pp. 697–732.

2 Helfferich, F., *Ion Exchange* (New York: McGraw-Hill, 1962).

3 Helfferich, F. and M. S. Plesset, *J. Chem. Phys.*, **28**, 418 (1958).

4 Montgomery, *Water Treatment Principles and Design* (Consulting Engineers, 1985).

5 Kunin, R., *Ion-Exchange Resins*, 2nd edn (New York: Wiley, 1958).

6 Wankat, P. C., *Rate Controlled Separations*, Chap. 9 (Kluwer Academic Publishers, 1990).

7 Reynolds, T. D., *Unit Operations and Processes in Environmental Engineering* (Boston: PWS Publishing Co., 1982).

8 Helfferich, F. and G. Klein, *Multicomponent Chromatography* (New York: Marcel-Dekker, 1970).

9 Klein, G., D. Tondeur, and T. Vermeulen, Multicomponent ion exchange in fixed beds, *I.E.C. Fund.*, **6**(3), 351–361 (1967).

10 Clifford, D. A. and W. J. Weber, Multi-component ion exchange: nitrate removal process with land disposal regenerant, *Ind. Water Eng.*, **15**(2), 18–26 (March, 1978).

11 Brignal, W. J., A. K. Gupta, and M. Streat, *The Theory and Practice of Ion Exchange* (Paper 11) (London: Society of Chemical Industry, 1976).

12 Smith, R. P. and E. T. Woodburn, *AIChE J.*, **24**, 577 (1978).

13 Robinson, R. A. and R. H. Stokes, *Electrolyte Solutions* (London: Butterworths, 1959).

14 Wilson, G. M., *J. Am. Chem. Soc.*, **86**, 127 (1964).

15 Schaeiwitz, J. A., written communications, 1997.

CHAPTER 9

1 Noble, R. D., An overview of membrane separations, *Sep. Sci. Tech.*, **22**, 731–743 (1987).

2 Mulder, M., *Basic Principles of Membrane Technology* (Boston: Kluwer Academic, 1991).

3 Koros, W. J., Membranes: learning lessons from nature, *Chem. Eng. Prog.*, **91**(10), October 1995.

4 Zolandz, R. R. and G. K. Fleming, Design of gas permeation systems, *Membrane Handbook*, Chap. 4, W. S. W. Ko and K. K. Sirkar, eds. (Chapman & Hall, 1992).

5 Seader, J. D. and E. J. Henley, *Separation Process Principles* (J. Wiley & Sons, Inc., 1998).

6 Hestekin, J. A., D. Bhattacharyya, S. K. Sikdar, and B. M. Kim, Membranes for treatment of hazardous wastewater, *Encyclopedia of Environmental Analysis and Remediation*, R. A. Meyers, ed. (J. Wiley & Sons, Inc., 1998).

7 Drioli, E., with permission.

8 Osmonics, Inc., with permission.

9 Bilstad, T., E. Espedal, A. H. Haaland, and M. Madland, Ultrafiltration of oily wastewater, *Second. Int. Conf. on Health, Safety & Environment in Oil & Gas Exploration Production*, Jakarta, Indonesia (1994).

10 http://www.wqa.org (current as of 19 June, 2003).

11 Reynolds, T. D., *Unit Operations and Processes in Environmental Engineering* (Boston: PWS Publishing, 1982).

12 Chapman-Wilbert, M., F. Leitz, E. Abart, B. Boegli, and K. Linton, *The Desalting and Water Treatment Membrane Manual: A Guide to Membranes for Municipal Water Treatment*, 2nd edn, Water Treatment Technology Program Report N. 29 (US Dept. of Interior, Bureau of Reclamation, July 1998).

13 *Membrane Separations Systems*, US Dept. of Energy Report DOE/ER/30133-H1,PS-1 (1990).

14 Baker, R. W. *et al.*, *Development of Pervaporation to Recover and Reuse Volatile Organic Compounds from Industrial Waste Streams*, US Dept. of Energy Final Report DOE/AL/98769-1 (DE97006846) March 1997.

15 Sherwood, T., P. Brian, and R. Fisher, Desalination by Reverse Osmosis, *Ind. Engg Chem. Fundamentals*, **6**, 2 (1967).

16 Gekas, V. and B. Hallstrom, Mass transfer in the membrane concentration polarization layer under turbulent cross-flow, *J. Mem. Sci.*, **30**, 153 (1987).

17 Davis, R. H., Cross-flow microfiltration with backpulsing, *Membrane Separations in Biotechnology*, 2nd edn, W. K. Wang, ed. (New York: Marcel-Dekker Pub. Co., 2000).

Index

Bold indicates chapters.